8th International Conference on New Computational Methods for Inverse Problems (NCMIP 2018)

Journal of Physics: Conference Series
Volume 1131

Cachan, France
25 May 2018

ISBN: 978-1-5108-7726-9
ISSN: 1742-6588

Printed from e-media with permission by:

Curran Associates, Inc.
57 Morehouse Lane
Red Hook, NY 12571

Some format issues inherent in the e-media version may also appear in this print version.

Institute of Physics
Dirac House, Temple Back
Bristol BS1 6BE UK

Phone: 44 1 17 929 7481
Fax: 44 1 17 920 0979

techtracking@iop.org

Additional copies of this publication are available from:

Curran Associates, Inc.
57 Morehouse Lane
Red Hook, NY 12571 USA
Phone: 845-758-0400
Fax: 845-758-2633
Email: curran@proceedings.com
Web: www.proceedings.com

Table of contents

Volume 1131

8th International Conference on New Computational Methods for Inverse Problems

25 May 2018, Cachan, France

Accepted papers received: 7 November 2018
Published online: 23 November 2018

Preface

Papers

IOP Conf. Series: Journal of Physics: Conf. Series **1131** (2018) 011001 doi:10.1088/1742-6596/1131/1/011001

Contents

Determination of blur kernel for HR-pQCT with bilevel optimization
Bruno Sixou, Yufei Li, Françoise Peyrin

Advanced statistical learning method for multi-physics NDT-NDE
Shamim Ahmed, Pierre Calmon, Roberto Miorelli, Christophe Reboud and Andrea Massa

A Bregman inexact linesearch–based forward–backward algorithm for nonsmooth nonconvex optimization
Simone Rebegoldi, Silvia Bonettini and Marco Prato

Multiple variational image assimilation for accessible micro-elastography
Aleix Boquet-Pujadas, Jean-Christophe Olivo-Marin

IOP Conf. Series: Journal of Physics: Conf. Series **1131** (2018) 011001 doi:10.1088/1742-6596/1131/1/011001

Foreword

This volume of Journal of Physics: Conference Series is dedicated to the scientific contributions presented during the 8th International Conference on New Computational Methods for Inverse Problems, NCMIP 2018 (http://complement.farman.ens-cachan.fr/NCMIP_2018.html). This conference took place in Cachan, France, at Ecole normale supérieure Paris-Saclay, on May 25, 2018.

The New Computational Methods for Inverse Problems (NCMIP) conference focused on recent advances in the resolution of inverse problems. Indeed, inverse problems appear in numerous scientific areas such as geophysics, biological and medical imaging, material and structure characterization, electrical, mechanical and civil engineering, and finances. The resolution of inverse problems consists in estimating the parameters of the observed system or structure from data collected by an instrumental sensing or imaging device. Its success firstly requires the collection of relevant observation data. It also requires accurate models describing the physical interactions between the instrumental device and the observed system, as well as the intrinsic properties of the solution itself. Finally, it requires the design of robust, accurate and efficient inversion algorithms. Advanced sensor arrays and imaging devices provide high rate and high volume data; in this context, the efficient resolution of the inverse problem requires the joint development of new models and inversion methods, taking computational and implementation aspects into account. During this one-day conference, researchers had the opportunity to bring to light and share new techniques and results in the field of inverse problems.

The topics of the conference were: algorithms and computational aspects of inversion, Bayesian estimation, Kernel methods, learning methods, convex optimization, free discontinuity problems, metamodels, proper orthogonal decomposition, reduced models for the inversion, non-linear inverse scattering, image reconstruction and restoration, and applications (bio-medical imaging, non-destructive evaluation...).

NCMIP 2018 was a one-day conference held in May 2018 which attracted around seventy attendees. Each of the submitted papers has been reviewed by two reviewers. There have been fourteen accepted papers. In addition, three international speakers were invited to present a longer talk.

The conference was supported by Institut Farman (ENS Paris-Saclay, CNRS) and endorsed by the following French research networks: GDR ISIS, GDR MIA, GDR MOA, GDR Ondes. The program committee acknowledges the following research laboratories: CMLA, LMT, LSV, LURPA and SATIE.

Eric Vourc'h

IOP Conf. Series: Journal of Physics: Conf. Series **1131** (2018) 011001 doi:10.1088/1742-6596/1131/1/011001

Conference co-chairs:

Eric Vourc'h, SATIE laboratory, ENS Paris-Saclay, CNRS, France
Thomas Rodet, SATIE laboratory, ENS Paris-Saclay, CNRS, France

Technical program committee:

Alexandre Baussard, ENSTA Bretagne, Lab-STICC, France
Silvia Bonettini, Università degli studi di Ferrara, dipartimento di matematica informatica, Ferrara, Italy.
Marc Bonnet, ENSTA, ParisTech, France
Laure Blanc-Féraud, I3S laboratory and INRIA Nice Sophia-Antipolis, CNRS, France
Antonin Chambolle, CMAP, Ecole Polytechnique, CNRS, France
Marion Darbas, LAMFA, CNRS UMR 7352, Université de Picardie Jules Verne, France
Oliver Dorn, School of Mathematics, University of Manchester, UK
Laurent Fribourg, LSV, ENS Paris-Saclay, CNRS, France
Jerôme Idier, IRCCyN Laboratory, Ecole Centrale de Nantes, France
Pierre-Yves Joubert, IEF, Paris-Sud University, CNRS, France
Marc Lambert, Geeps Laboratory, CNRS, CentraleSupElec, Paris-Sud University, France
Giacomo Oliveri, eledia research center/eledia@L2S group, University of Trento, Italy
Dominique Lesselier, L2S Laboratory, CNRS, CentraleSupElec, Paris-Sud University, France
Matteo Pastorino, DIBE, University of Genoa, Italy
Gabriel Peyré, DMA, ENS, CNRS, France
Anthony Quinn, Trinity College Dublin, Ireland
Marco Salucci, eledia research center/eledia@L2S group, University of Trento, Italy

Local organizing committee chair:

Eric Vourc'h, SATIE laboratory, ENS Paris-Saclay, Paris-Saclay University, CNRS, France

Local organizing committee:

Virginie Guénard, LSV Laboratory, ENS Paris-Saclay, CNRS, France
Virginie Pauchont, CMLA Laboratory, ENS Paris-Saclay, CNRS, France
Sophie Abriet, SATIE Laboratory, ENS Paris-Saclay, CNRS, France
Béatrice Bacquet, SATIE Laboratory, ENS Paris-Saclay, CNRS, France
Dominique Bach, SATIE Laboratory, ENS Paris-Saclay, , France

Invited speakers

Gabriele Steidl, Technische Universität Kaiserslautern, Fachbereich Mathematik, Germany
Anne Sentenac, Institut Fresnel, CNRS, Marseille, France
Philippe Ciuciu, Neurospin, CEA Saclay, France

IOP Conf. Series: Journal of Physics: Conf. Series **1131** (2018) 011002 doi:10.1088/1742-6596/1131/1/011002

Peer review statement

All papers published in this volume of *Journal of Physics: Conference Series* have been peer reviewed through processes administered by the proceedings Editors. Reviews were conducted by expert referees to the professional and scientific standards expected of a proceedings journal published by IOP Publishing.

IOP Conf. Series: Journal of Physics: Conf. Series **1131** (2018) 012001 doi:10.1088/1742-6596/1131/1/012001

A level set method for shape reconstruction in seismic full waveform inversion using a linear elastic model in 2D

Yifan Wu and Oliver Dorn

School of Mathematics, The University of Manchester, UK

E-mail: oliver.dorn@manchester.ac.uk

Abstract. We present a novel shape reconstruction technique for seismic full waveform inversion that uses a linear elastic system in 2D for modeling wave propagation in the Earth and a level set technique for performing shape evolution. A cost functional is formulated and minimized by a gradient based descent technique with backtracking line search and an Armijo condition. The three distributed elastic parameter profiles are in our model assumed to follow a given reference profile inside an unknown region D, and a different reference profile outside of D. The task is to reconstruct the unknown region D from seismic data obtained at or close to the surface. We present numerical experiments that address the imaging of salt domes buried in the Earth. The results demonstrate that our new algorithm is able to reconstruct shapes with non-trivial topology from a simpler starting guess in few iterations.

1. Introduction

Seismic full waveform inversion (FWI) has become an interesting alternative to more traditional geophysical exploration techniques [12]. A sufficiently accurate model for describing seismic wave propagation in the Earth is given by a linear elastic system, which however is high-dimensional and therefore difficult to handle computationally as well as theoretically. Due to this, many practical algorithms for data inversion are based on lower-dimensional approximations such as the acoustic wave equation in time domain or a Helmholtz equation in frequency domain. However, it is not clear whether those simplified models are able to correctly represent fundamental physical processes occurring during the wave propagation, possibly leading to incorrect results or other difficulties. This is why in our study we use a linearised elastic wave model instead which is assumed to be more accurate. So far this is done in a proof-of-concept approach using a 2D setup. We anticipate, however, that the algorithm as well as general observations directly extend to a more realistic elastic 3D setup.

Regardless of the model used for wave propagation, the inverse problem of seismic FWI is highly ill-posed such that strong regularization techniques are necessary in order to achieve stable results. Usually regularization techniques aim at incorporating any available a-priori information on the desired solution into the reconstruction. In the work presented here, we use a shape-based approach for this task which is targeting situations where structural a-priori information is available from additional imaging modalities such as gravity, well-logs, controlled source electromagnetic or outcrop data which often can be carried out prior to a full-waveform seismic survey.

During the iterative non-linear inversion strategy, an initial profile is successively modified until a provided cost functional is minimized. We use a level set representation of shapes for this iteration which provides us with sufficient flexibility to adjust the interfaces during the reconstruction process to the unknown topology of the hidden objects in an efficient and automatic way. Related approaches for a simpler scalar acoustic or Helmholtz model have been presented in [3, 4, 5]. Our approach goes beyond that work in that it directly incorporates a more complex time-dependent elastic wave propagation model in the level set evolution.

2. The elastic forward model for waveform inversion

We model seismic wave propagation in a 2D Earth by a first-order symmetric hyperbolic system [7] of the following form

$$\begin{aligned}
\frac{\partial v_x}{\partial t} &= \frac{1}{\rho}\left(\frac{\partial \eta_{xx}}{\partial x} + \frac{\partial p}{\partial x} + \frac{\partial \eta_{xy}}{\partial y} + F_x\right) \\
\frac{\partial v_y}{\partial t} &= \frac{1}{\rho}\left(\frac{\partial \eta_{xy}}{\partial x} + \frac{\partial p}{\partial y} + \frac{\partial \eta_{yy}}{\partial y} + F_y\right) \\
\frac{\partial \eta_{xx}}{\partial t} &= 2\mu \frac{\partial v_x}{\partial x} \\
\frac{\partial \eta_{yy}}{\partial t} &= 2\mu \frac{\partial v_y}{\partial y} \\
\frac{\partial \eta_{xy}}{\partial t} &= \mu\left(\frac{\partial v_y}{\partial x} + \frac{\partial v_x}{\partial y}\right) \\
\frac{\partial p}{\partial t} &= \lambda\left(\frac{\partial v_x}{\partial x} + \frac{\partial v_y}{\partial y}\right)
\end{aligned} \tag{1}$$

with appropriate initial and boundary conditions. It models elastic wave propagation inside a domain Ω [7] where ρ denotes density and μ and λ are two Lamé parameters. The quantity $\boldsymbol{v} = (v_x, v_y)^T$ is a vector function describing velocity, η is a 2×2 tensor function related to the stress tensor, and p is a scalar function describing pressure. F_x and F_y are seismic source terms, t denotes time and x and y the two cartesian directions.

The unknown model parameters (ρ, μ, λ) used in (1) are related to the P and S wave velocities V_p and V_s by

$$V_p = \sqrt{\frac{\lambda + 2\mu}{\rho}}, \qquad V_s = \sqrt{\frac{\mu}{\rho}}, \tag{2}$$

which gives rise to the equivalent alternative representation of model parameters as (ρ, V_p, V_s). We will use both representations in this paper. The system (1) can be written in more compact notation as

$$\mathcal{L}(\boldsymbol{m})\boldsymbol{w} = \Gamma(\boldsymbol{m})\frac{\partial \boldsymbol{w}}{\partial t} + D_x \frac{\partial \boldsymbol{w}}{\partial x} + D_y \frac{\partial \boldsymbol{w}}{\partial y} = \boldsymbol{q} \tag{3}$$

where $\boldsymbol{m}$ represents the estimated model parameters (ρ, μ, λ), $\boldsymbol{w}$ the states, $\boldsymbol{q}$ the seismic source, $\Gamma(\boldsymbol{m})$ is a diagonal matrix containing functions of (ρ, μ, λ) on its diagonal, and D_x, D_y are scalar symmetric matrices.

Let us assume that we apply $s = 1, \ldots, S$ different source patterns $\boldsymbol{q}^s$ (usually located at discrete positions $\boldsymbol{x}_s$ with a prescribed time dependence) giving rise to solutions $\boldsymbol{w}^s$ of (3). For each source, we measure these states $\boldsymbol{w}^s$ at a discrete set of $r = 1, \ldots, R$ different receivers located at positions $\boldsymbol{x}_r$. In our case, source positions $\boldsymbol{x}_s$ and receiver positions $\boldsymbol{x}_r$ are close to the surface above the region of interest. Let us denote these measured data by $\widehat{\boldsymbol{d}}_{sr} = \mathcal{P}_r \boldsymbol{w}^s$ where $\mathcal{P}_r$ is the linear measurement operator at receiver with index r. For example, it might record point measurements of all components of $\boldsymbol{w}^s$ at the location $\boldsymbol{x}_r$. For a given model $\boldsymbol{m}$ we can simulate corresponding data by running our forward model with a seismic simulator producing simulated data $\boldsymbol{d}_{sr}[\boldsymbol{m}]$. The mathematical task in the inverse problem will be to minimize the mismatch between predicted data $\boldsymbol{d}_{sr}[\boldsymbol{m}]$ and measured data $\widehat{\boldsymbol{d}}_{sr}$ in some sense. Here we use the minimization of a classical least squares data misfit for that purpose, namely

$$min_{\boldsymbol{m}} E(\boldsymbol{m}) = \frac{1}{2}\sum_{s=1}^{S}\sum_{r=1}^{R} \|\boldsymbol{d}_{sr}[\boldsymbol{m}] - \widehat{\boldsymbol{d}}_{sr}\|_2^2 \tag{4}$$

constrained by

$$\mathcal{P}_r \boldsymbol{w}^s = \boldsymbol{d}_{sr}[\boldsymbol{m}] \quad (\textit{State-data map})$$

$$\mathcal{L}(\boldsymbol{m})\boldsymbol{w}^s = \boldsymbol{q}^s \quad (\textit{Elastic wave equation}).$$

This inverse problem is ill-posed and requires some form of regularization for its stable solution. We are using a model based approach for this regularization task based on a level set representation of parameters as explained in the next section.

3. The new level set shape reconstruction approach

In this work we assume that the unknown model parameters are shape-like in the sense that regions filled with constant and known parameters are embedded in a smoothly varying (also known) background. This way we reduce the number of unknowns and thereby stabilize the underlying inverse problem. Furthermore, we assume that all model parameters share the same regions and interfaces towards the smooth background profile. The main application we have in mind is the mapping of salt bodies in seismic full waveform inversion. A similar approach has been taken in [3, 4, 5], however using a scalar Helmholtz model in frequency domain instead.

In our level set approach, we assume that the parameters ρ, μ and λ obtain a prescribed functional profile $\boldsymbol{m}_{obj}(\boldsymbol{x})$ inside an unknown region D (which can have complicated shape) and a different functional profile $\boldsymbol{m}_{back}(\boldsymbol{x})$ outside of D. As already mentioned, in this paper we assume in particular that $\boldsymbol{m}_{obj}(\boldsymbol{x})$ assumes constant values for all model parameters. This assumption will be relaxed in future work, but is a good approximation when mapping salt bodies in the geophysical prospecting of hydrocarbon reservoirs.

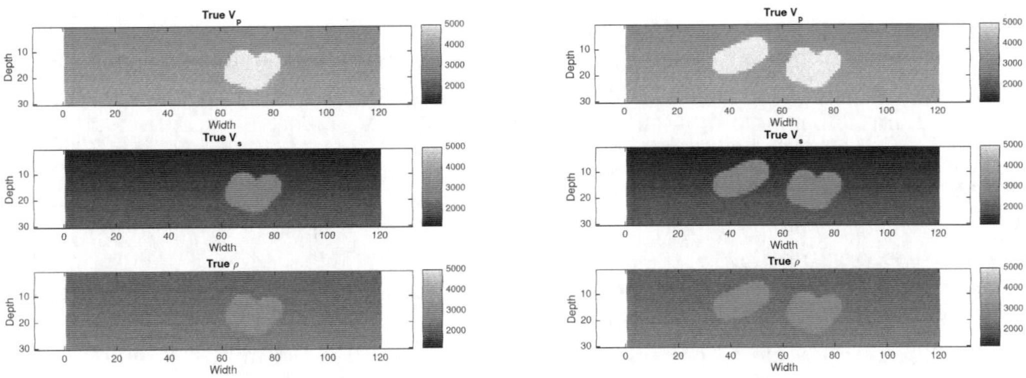

Figure 1. Reference models for our numerical experiments. Left: test case 1; right: test case 2. For both the upper image indicates V_p, the central image V_s and the bottom image ρ.

IOP Conf. Series: Journal of Physics: Conf. Series **1131** (2018) 012001 doi:10.1088/1742-6596/1131/1/012001

In the level set representation employed here, the unknown region D is implicitly represented by a level set function ϕ according to

$$D = \{\boldsymbol{x} \,|\, \phi(\boldsymbol{x}) \geq 0\}\,,$$
$$\Omega \backslash D = \{\boldsymbol{x} \,|\, \phi(\boldsymbol{x}) < 0\}\,.$$

The model parameter $\boldsymbol{m}$ is then parameterised as

$$\boldsymbol{m} = \boldsymbol{m}_{obj}(\boldsymbol{x}) \quad \text{for} \quad \boldsymbol{x} \in D\,,$$
$$\boldsymbol{m} = \boldsymbol{m}_{back}(\boldsymbol{x}) \quad \text{for} \quad \boldsymbol{x} \in \Omega \backslash D\,.$$

For details of the level set technique see for example [1, 6, 10]. Using the one-dimensional Heaviside function this can alternatively be written as

$$\boldsymbol{m} = \boldsymbol{m}_{obj}(\boldsymbol{x}) H(\phi(\boldsymbol{x})) + \boldsymbol{m}_{back}(\boldsymbol{x})(1 - H(\phi(\boldsymbol{x}))\,. \tag{5}$$

In the numerical simulations presented further below we assume that $\boldsymbol{m}_{obj}(\boldsymbol{x})$ is equal to a given constant value representing salt properties, and $\boldsymbol{m}_{back}(\boldsymbol{x})$ represents a typical background profile which is assumed to approximately follow a linear dependence on depth and being horizontally invariant.

A standard way of minimizing the least squares data functional (4) in a pixel or voxel based formulation for the parameters is to employ gradient techniques by a reverse-time migration adjoint-state method [2, 8]. We follow a similar route, using however the above formulation of a symmetric hyperbolic system (3) for wave propagation, and furthermore extend the resulting expressions to our level set formulation. In this new approach, gradients need to be expressed with respect to the level set function, which is achieved for example by formally applying the chain rule to the results of the classical adjoint technique using (5).

Assume that we have computed the gradient $\boldsymbol{g_m} = \partial E / \partial \boldsymbol{m}$ following such a modification of the adjoint-state method. By formally applying the chain rule, we obtain the level set based gradient $\boldsymbol{g}_\phi = \partial E / \partial \phi$ satisfying

$$\boldsymbol{g}_\phi = [\boldsymbol{g}_{\boldsymbol{m}}^T (\boldsymbol{m}_{obj} - \boldsymbol{m}_{back})] \delta(\phi(\boldsymbol{x})) \tag{6}$$

where ^{T} means 'transpose'. Once this gradient $\boldsymbol{g}_\phi$ has been calculated, we will follow here a steepest descent style approach (using $\boldsymbol{g}_\phi$) on the level set function combined with a backtracking line search with Armijo condition and a narrow-band approximation to $\delta(\phi(\boldsymbol{x}))$ in each iteration. The numerical implementation of this approach results in a shape evolution which, upon convergence, will deliver a shape that minimizes the least squares data misfit functional (4). For more details we refer here to [1, 9].

4. Numerical experiments

We test our algorithm by using two different reference setups displayed in Figure 1, both addressing the imaging of salt domes buried in the ground with FWI. The left hand column of the figure shows a profile where only one salt dome is included in the ground. The right hand column shows a slightly more complicated setup with two separate salt domes hidden in the ground. The top row of this figure shows the p-wave velocities V_p, the center row the s-wave velocities V_s, and the bottom row the densities ρ for both cases. Notice that this figure is displayed with equal axes in x- and y-directions, which indicates the real physical dimensions of the field. All following images presented here will use different axes for x- and y-directions in order to obtain a more compact form of visualization. We use 21 equidistant seismic point sources distributed along a horizontal line close to the surface (i.e. close to the top of the computational domain displayed in Figure 1) with time dependence being of Ricker wavelet type in order to generate our data. Receivers are located at 81 equidistant locations along a different horizontal line close to the surface measuring the time series of all components of the arriving elastic waves at the receiver locations. This means that we have only a top view on the domain of interest available for gathering data, as it is typical in seismic surveys.

As forward solver for generating synthetic data as well as for reconstruction we use the elastic wave propagation module of the k-Wave toolbox [11]. We discretize the domain of interest into a grid of size 30×120 pixels each having dimension $100\text{m} \times 100\text{m}$. Therefore the total physical size of the test domain is 3km $\times$ 12km. The parameter values in each region can be seen in Figure 1. In this proof-of-concept study, 5% Gaussian noise is added to the data prior to running the reconstruction algorithm, in order to avoid the inverse crime. In future work, different discretization levels for simulating data will be tested as well.

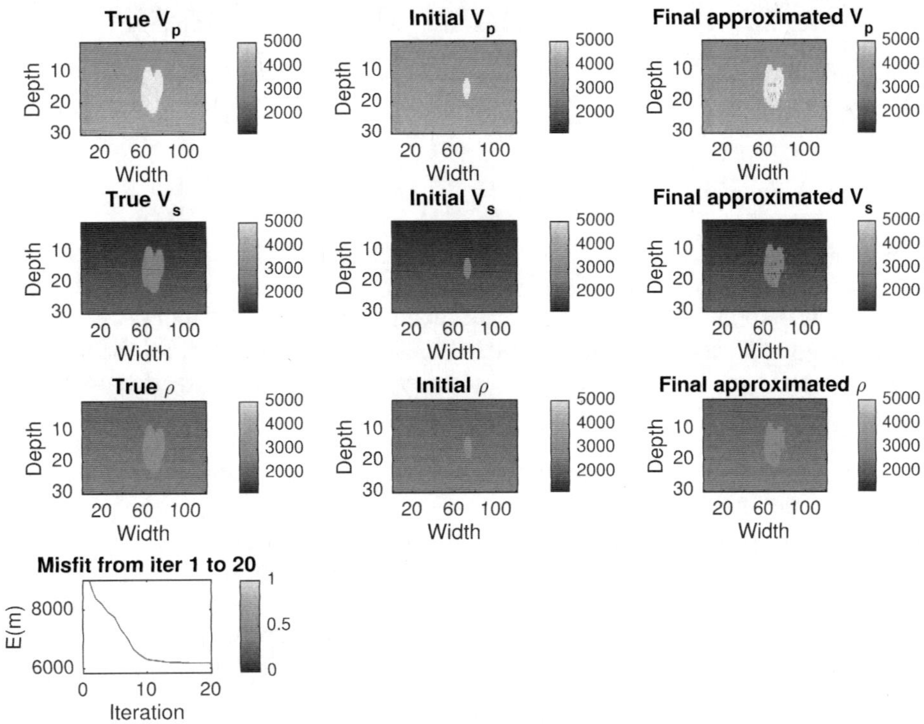

Figure 2. Level set shape reconstruction for the first test case. Left column: reference model as on the left of Figure 1; central column: initial guess; right column: final reconstruction. The bottom image shows the evolution of the least squares data misfit (4) against iteration number.

Figure 2 demonstrates the result of the first test case with just one embedded object after running the algorithm for 20 iterations. As stopping criterion for both cases we assume that the relative error for the data misfit is less than 10^{-6}. The top row shows V_p, the center row V_s and the bottom row ρ for each case. More details are provided in the caption of the figure.

Figure 3 demonstrates the results of the second test case with the same arrangement of images as in Figure 2. In this slightly more complicated geometry, 24 iterations are needed to obtain convergence.

Figure 4 visualizes the shape evolution for the second test case in steps of 3 iterations using

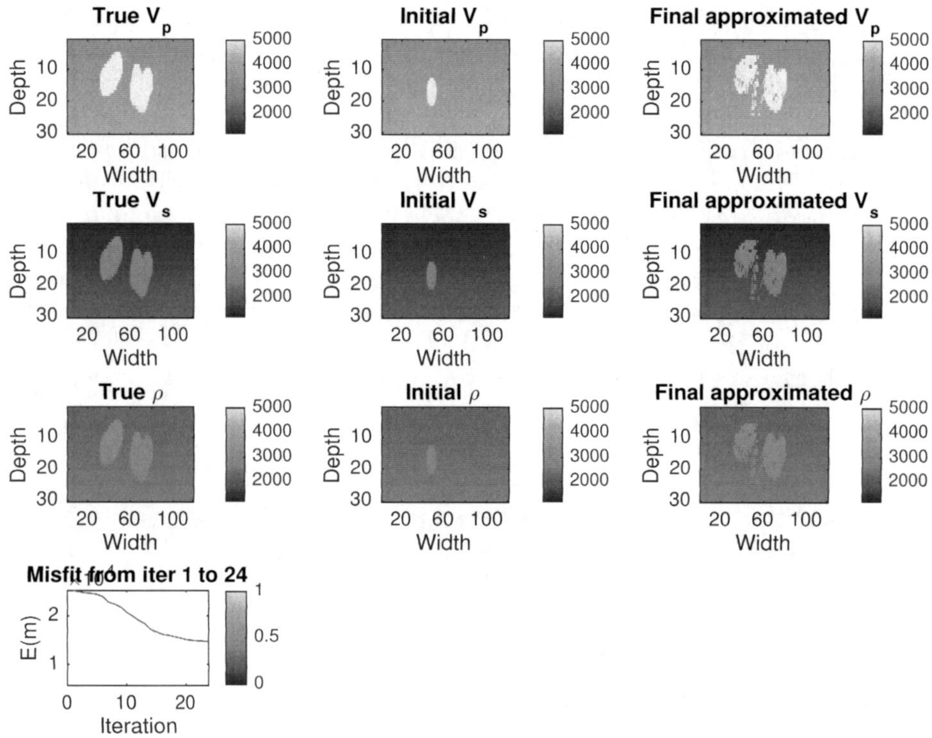

Figure 3. Level set shape reconstruction for the second test case. Left column: reference model as on the right of Figure 1; central column: initial guess; right column: final reconstruction. The bottom image shows the evolution of the least squares data misfit (4) against iteration number.

V_p as reference profile. The shape evolution of the V_s and ρ profiles is similar since they all share the same level set function.

We observe that in both cases the algorithm is able to obtain a very good approximation of the correct shapes of salt domes in the area of interest. The obtained shapes appear slightly irregular. This could be controlled, if desired, by either adding specific regularization terms to (4) or by using Sobolev gradients for the level set update as in [9]. In deeper areas some artefacts build up in the second test case which as well might be avoidable when applying stronger regularization tools.

5. Summary and future research

We have proposed a novel shape-based reconstruction technique for seismic FWI using a linear elastic system for the forward modelling and a level set technique for shape evolution. A gradient based line search strategy provides an efficient reconstruction which converges to a shape that approximates the reference model well. Therefore, this novel technique represents a promising alternative to more traditional pixel or voxel based reconstruction strategies as long as the available a-priori information justifies the incorporated structural information.

Future work will include the development of additional regularization techniques for controlling artefacts in deeper regions and obtaining smoother shape boundaries. Also the simultaneous estimation of internal and external profiles with the correct shapes, and the separate estimation of parameter-specific region for V_p, V_s and ρ, are challenging tasks to work on. In addition, more efficient minimization approaches for the cost functional (4) will be investigated as for example a nonlinear Kaczmarz type approach. This includes the investigation

IOP Conf. Series: Journal of Physics: Conf. Series **1131** (2018) 012001 doi:10.1088/1742-6596/1131/1/012001

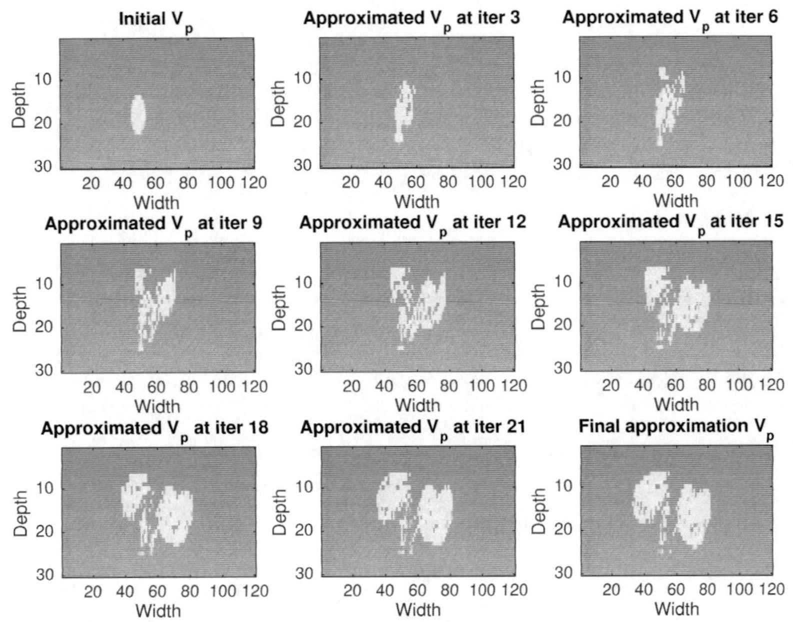

Figure 4. Shape evolution for the second test case: shown is V_p for different iterations.

of optimal data types, measurement geometries and suitable stopping criteria.

References

[1] O. Dorn and D. Lesselier, *Level set methods for inverse scattering*, Inverse Problems, Volume 22, Number 4, 2006.

[2] O. Dorn, *Time-reversal and the adjoint imaging method with an application in telecommunication*, Inverse problems and imaging, Volume 1943 of the series Lecture Notes in Mathematics, pp 135-170, 2008.

[3] Z. Guo, M. De Hoop, *Shape optimization and level set method in full waveform inversion with 3D body reconstruction*, SEG Houston Annual Meeting, Technical Program Expanded Abstracts 2013: pp. 1079-1083.

[4] A. Kadu, T. van Leeuwen and W.A. Mulder, *Salt reconstruction in full waveform inversion with a parametric level-set method*, IEEE Trans. Computational Imaging, Vol 3 (2), pp 305–315, 2017.

[5] W. Lewis, B. Starr, and D. Vigh, *A level set approach to salt geometry inversion in full waveform inversion*, SEG Las Vegas 2012 Annual Meeting, 2012.

[6] S. Osher and J. Sethian, *Fronts propagating with curvature dependent speed: Algorithms based on Hamilton-Jacobi formulations*, Journal of Computational Physics, Vol 56, pp 12–49, 1988.

[7] G. Papanicolaou, L. Ryzhik, and J. Keller, *Transport equations for Elastic and Other Waves in Random Media*, Wave Motion, Volume 24, Issue 4, Pages 327-370, December, 1996.

[8] R. Plessix, *A review of the adjoint-state method for computing the gradient of a functional with geophysical applications*, Geophysical Journal, Volume 167, Issue 2, 495-503, 2006.

[9] K. Prieto and O. Dorn, *Sparsity and level set regularization for diffuse optical tomography using a transport model in 2D*, Inverse Problems, **33** (1) 2016.

[10] F. Santosa, *A level set approach for inverse problems involving obstacles*, ESAIM: Control, Optimization and Calculus of Variations, Vol 1, pp 17-33, 1996.

[11] B.E. Treeby, J. Jaros, D. Rohrbach and B.T. Cox, *Modelling elastic wave propagation using the k-Wave MATLAB toolbox*, IEEE International Ultrasonics Symposium, pp. 146–149, 2014.

[12] J. Virieux, S. Operto, *An overview of full-waveform inversion in exploration geophysics*, Geophysics, VOL.74, No.6, 2009.

IOP Conf. Series: Journal of Physics: Conf. Series **1131** (2018) 012002 doi:10.1088/1742-6596/1131/1/012002

Instance Optimal Decoding and the Restricted Isometry Property

Nicolas Keriven

École Normale Supérieure. 45 rue d'Ulm, 75005 Paris, France.

E-mail: `nicolas.keriven@ens.fr`

Rémi Gribonval

Université Rennes 1, Inria, CNRS, IRISA. F-35000 Rennes, France.

E-mail: `remi.gribonval@inria.fr`

Abstract. In this paper, we address the question of *information preservation* in ill-posed, non-linear inverse problems, assuming that the measured data is close to a low-dimensional model set. We provide necessary and sufficient conditions for the existence of a so-called *instance optimal* decoder, *i.e.*, that is robust to noise and modelling error. Inspired by existing results in compressive sensing, our analysis is based on a (Lower) *Restricted Isometry Property* (LRIP), formulated in a non-linear fashion. We also provide sufficient conditions for non-uniform recovery with random measurement operators, with a new formulation of the LRIP. We finish by describing typical strategies to prove the LRIP in both linear and non-linear cases, and illustrate our results by studying the invertibility of a one-layer neural net with random weights.

1. Introduction

Inverse problems are ubiquitous in all areas of data science. While *linear* inverse problems have been arguably far more studied in the literature, some frameworks are intrinsically non-linear [15]. In this paper, we aim at giving a characterization of the *preservation of information* in ill-posed inverse problems, regularized by the introduction of a "low-dimensional" *model set* close to which the data of interest is assumed to live. We consider a very general context that includes, in particular, measurement operators that are possibly non-linear, and an ambient space that can be any pseudometric set. Our main results show that the existence of a decoder that is robust to noise and modelling error is *equivalent* to a modified *Restricted Isometry Property* (RIP), which is a classical property in compressive sensing [9]. We thus outline the fundamental nature of the RIP in settings that are more general than previously studied.

The problem is formulated as follows. Let (E, d_E) be a set equipped with a pseudometric[1] d_E, the set of data, and $(F, \|\cdot\|_F)$ a seminormed[2] vector space, the space of measurements. Consider a (possibly non-linear) measurement map $\Psi : E \to F$. The measured vector is:

$$\mathbf{y} = \Psi(\mathbf{x}^\star) + \mathbf{e} \tag{1}$$

[1] A pseudometric d satisfies all the requirements of a metric except $d(x, y) = 0 \Rightarrow x = y$.

[2] Similarly, a seminorm satisfy the requirements of a norm except that $\|x\| = 0$ does not imply $x = 0$.

where $\mathbf{e} \in F$ is measurement noise and $\mathbf{x}^\star$ is the true signal. Our goal is to characterize the existence of *any* procedure that would allow us to approximately recover the data $\mathbf{x}^\star$ from $\mathbf{y}$.

Regularization. In most interesting problems, the "dimension" of the space F is far lower than that of the set E (in a loose sense: we recall that here none is required to be finite-dimensional, and E is not necessarily a vector space), which makes the problem *ill-posed*, meaning that there are information-theoretic limits that prevent us from recovering the underlying signal from the measurements, even in the noiseless case. A classical regularization technique is to introduce *prior knowledge* about the true signal $\mathbf{x}^\star$, here we consider a *model set* $\mathfrak{S} \subset E$ of "simple" signals, such that $\mathbf{x}^\star$ is likely to be close to $\mathfrak{S}$. For instance, sparsity, *i.e.* the assumption that the true signal is a linear combination of a few elements in a well-chosen dictionary, is a hugely-studied prior in modern signal processing, in particular in compressive sensing [16].

Instance Optimal Decoding. Ideally, a *decoder* $\Delta : F \to E$ must be able to exactly retrieve $\mathbf{x}^\star$ from $\mathbf{y}$ when the modelling is exact (*i.e.* $\mathbf{x}^\star \in \mathfrak{S}$) and the noise is zero ($\mathbf{e} = 0$). However, as these conditions are highly unrealistic in practice, it is desirable for this decoding process to be both robust to noise and stable to *modelling error*. In the literature, such a decoder is said to be *instance optimal* [13] (see Def. 1). In this paper, our goal is to characterize the *existence* of an instance optimal decoder for the problem (1). In particular, we will not study the existence of efficient algorithms to solve (1), which is another significant achievement of compressive sensing [16], but only the *preservation of information* of the encoding process.

In [13, 8], the authors outlined the crucial role played by the *Restricted Isometry Property* (RIP), and more precisely by the *Lower*-RIP (LRIP), for the existence of instance optimal decoders in the linear case. In this paper, we extend these results to the non-linear case and to non-uniform probabilistic recovery.

Outline of the paper. The structure of the paper is as follows. In Section 2 we briefly outline some relevant references, keeping in mind that the field is large and we do not pretend to be exhaustive in this short paper. In Section 3 we state our main results relating instance optimal decoders (Def. 1) and the LRIP (Def. 2). In Section 4 we outline how one might typically prove the LRIP by extending a classical proof [2], and illustrate it on a simple example.

2. Related Work

Classical Compressive Sensing: the linear case. Instance optimal decoding and the RIP are well-known notions in compressive sensing [12, 11, 14]. We refer to the book of Foucart and Rauhut [16] for a review of the field, in particular to Chapters 6 and 11. The interplay between the two notions was in particular studied in [13] in the finite-dimensional case. These results were later extended to more general models in [22], and to any linear measurement operators in [8]. Of course, although these works and the present paper show that the LRIP is *necessary* for the existence of an instance optimal decoder, many works provide recovery guarantees (possibly weaker than instance optimality) without explicitely proving the RIP [10].

Non-uniform decoding. In compressed sensing the measurement operator Ψ is often designed at random. Typical recovery results are therefore given *with high probability.* One can then distinguished between *uniform* guarantees, meaning that with high probability on the draw of Ψ all signals $\mathbf{x}^\star$ close to $\mathfrak{S}$ can be stably recovered, and non-uniform guarantees, *i.e.* for one *fixed* signal $\mathbf{x}^\star$ close to $\mathfrak{S}$, with high probability on Ψ the decoding is successful. In [13] the authors study non-uniform instance optimality, but only under the light of the classical uniform RIP. In this paper we introduce a non-uniform version of the LRIP and prove that it is sufficient for non-uniform instance optimality.

IOP Conf. Series: Journal of Physics: Conf. Series **1131** (2018) 012002 doi:10.1088/1742-6596/1131/1/012002

Non-linear inverse problems. Non-linear inverse problems can be found in many areas of signal processing, see e.g. [15] for a review of some applications. They have also been considered by the compressive sensing community, often when *quantization* occurs [20, 7], in the so-called "1-bit" compressed sensing line of work [6]. Another focus is the development of efficient algorithms inspired by the linear case [3, 4]. In [4], the author assume that a locally linearized version of Ψ satisfy the classical RIP. In this paper we consider a different, "fully" non-linear RIP. We note that one notion does not imply the other.

3. Equivalence between IOP and LRIP

In this section we state our main results on instance optimal decoders and the LRIP. We distinguish the case where the operator Ψ is deterministic, or, equivalently, when it is random but one seeks so-called *uniform* recovery guarantees, and the case of non-uniform recovery.

3.1. Deterministic operator

Recall that we consider a pseudometric set (E, d_E), a seminormed vector space $(F, \|\cdot\|_F)$, and measurements of the form $\mathbf{y} = \Psi(\mathbf{x}^\star) + \mathbf{e}$ where $\Psi : E \to F$. We consider a model set $\mathfrak{S} \subset E$, and are interested in characterizing the existence of a good *decoder* $\Delta(\Psi, \mathbf{y})$ that takes Ψ and $\mathbf{y}$ as inputs and return a signal $\widetilde{\mathbf{x}} \in E$ that is close to $\mathbf{x}^\star$. We want this decoder to be stable to modelling error and robust to noise, which is characterized by the notion of instance optimality.

Definition 1 (Instance Optimality Property (IOP)). *A decoder Δ satisfies the Instance Optimality Property for the operator Ψ and model $\mathfrak{S}$ with constants $A, B > 0$, pseudometrics d_E, d'_E on E and error $\lambda \geq 0$ if: for all signals $\mathbf{x}^\star \in E$ and noise $\mathbf{e} \in F$, denoting $\widetilde{\mathbf{x}} = \Delta(\Psi, \Psi(\mathbf{x}^\star) + \mathbf{e})$ the recovered signal, it holds that:*

$$d_E(\mathbf{x}^\star, \widetilde{\mathbf{x}}) \leq A d'_E(\mathbf{x}^\star, \mathfrak{S}) + B \|\mathbf{e}\|_F + \lambda \tag{2}$$

where $d'_E(\mathbf{x}, \mathfrak{S}) = \inf_{\mathbf{x}_\mathfrak{S} \in \mathfrak{S}} d'_E(\mathbf{x}, \mathbf{x}_\mathfrak{S})$.

As indicated by the r.h.s. of (2), the decoding error between the true signal and the recovered one is bounded by the amplitude of the noise and the distance from $\mathbf{x}^\star$ to the model set, which indicates how well $\mathbf{x}^\star$ is modelled by $\mathfrak{S}$. An instance optimal decoder is therefore robust to noise and stable even if $\mathbf{x}^\star$ is not exactly in the model set. We also include a possible fixed additional error $\lambda \geq 0$, which may be unavoidable in some cases (due to algorithmic precision for instance). Ideally, one has $\lambda = 0$.

Let us now turn to the proposed non-linear version of the LRIP. As described in [8], the LRIP is just one side of the classical RIP, which states that the measurement operator Ψ approximately preserves distances between elements of the model $\mathfrak{S}$.

Definition 2 (Lower Restricted Isometry Property (LRIP)). *The operator Ψ satisfies the Lower Restricted Isometry Property for the model $\mathfrak{S}$ with constant $\alpha > 0$, pseudometric d_E and error $\eta \geq 0$ if: for all $\mathbf{x}, \mathbf{x}' \in \mathfrak{S}$ it holds that*

$$d_E(\mathbf{x}, \mathbf{x}') \leq \alpha \left\|\Psi(\mathbf{x}) - \Psi(\mathbf{x}')\right\|_F + \eta. \tag{3}$$

The LRIP expresses the fact that Ψ must not collapse two elements of the model together. Like the IOP, we allow for a possible additional fixed error $\eta \geq 0$ in the LRIP. Note that this type of error is often considered when introducing quantization [7, 20]. Ideally, one has $\eta = 0$, however in some cases it can be considerably simpler to prove that the LRIP holds with a non-zero η [21]. The reader would note that the classical RIP is often expressed with a constant $\alpha = (1-t)^{-1}$ where $t < 1$ is a small as possible.

We now state our main result. The proof, rather direct, can be found in Appendix A.

IOP Conf. Series: Journal of Physics: Conf. Series **1131** (2018) 012002 doi:10.1088/1742-6596/1131/1/012002

Theorem 1 (Equivalence between IOP and LRIP.). *Consider an operator Ψ and a model $\mathfrak{S}$.*

(i) If there exists a decoder Δ which satisfies the Instance Optimality Property for Ψ and $\mathfrak{S}$ with constants $A, B > 0$, pseudometrics d_E, d'_E and error $\lambda \geq 0$, then the operator Ψ satisfies the LRIP for $\mathfrak{S}$ with constant $\alpha = B$, pseudometric d_E and error $\eta = 2\lambda$.

(ii) If the operator Ψ satisfies the LRIP for the model $\mathfrak{S}$ with constant α, pseudometric d_E and error $\eta \geq 0$, then the decoder Δ defined as[3]

$$\Delta(\Psi, \mathbf{y}) = \arg\min_{\mathbf{x}\in\mathfrak{S}} \|\Psi(\mathbf{x}) - \mathbf{y}\|_F \tag{4}$$

satisfies the Instance Optimality Property for the operator Ψ and model $\mathfrak{S}$ with constants $A = 1$ and $B = 2\alpha$, pseudometrics d_E and d'_E where d'_E is defined by $d'_E(\mathbf{x}, \mathbf{x}') := d_E(\mathbf{x}, \mathbf{x}') + 2\alpha \|\Psi(\mathbf{x}) - \Psi(\mathbf{x}')\|_F$, and error $\lambda = \eta$.

Theorem 1 states that if the LRIP is satisfied, then the decoder that returns the element in the model that best matches the measurement is instance optimal, with a special metric d'_E. On the other hand, if *some* instance optimal decoder exists, then the LRIP must be satisfied. In other words, when the LRIP is satisfied, then we know that a negligible amount of information is lost when encoding a signal well-modeled by $\mathfrak{S}$. Conversely, if the LRIP is not satisfied, one has no hope of deriving an instance optimal decoder.

3.2. Random operator, from uniform recovery to non-uniform recovery

In the vast majority of the compressive sensing literature, the measurement process is drawn at random: for instance, in the finite dimensional case, it is an open problem to find deterministic matrices that satisfies the RIP with an optimal number of measurements ([16], pp. 27), while on the contrary many classes of random matrices satisfy the RIP with high probability [2].

A well-studied concept is that of uniform recovery guarantees, where one shows that, with high probability on the draw of Ψ, the LRIP holds. It follows by Theorem 1 that there is a decoder such that, with high probability on the draw of Ψ, all signals from $\mathfrak{S}$ can be stably recovered. There is also a notion of *non-uniform* recovery, where one considers a decoder Δ and wonders if, given an arbitrary signal close to $\mathfrak{S}$, this signal is stably recovered (with high probability on the draw of Ψ) from $\Psi(\mathbf{x}^\star) + \mathbf{e}$. In this section we introduce a non-uniform version of the LRIP, and show that it is a sufficient condition for the existence of a non-uniform instance optimal decoder. We start by discussing a notion of projection on the model.

Remark 1 (Approximate projection.). *As we will see, in non-uniform recovery the distance from $\mathbf{x}^\star$ to $\mathfrak{S}$ is replaced by the distance from $\mathbf{x}^\star$ to a particular element $\mathbf{x}_\mathfrak{S} = \mathrm{P}_\mathfrak{S}(\mathbf{x}^\star)$, where $\mathrm{P}_\mathfrak{S} : E \to \mathfrak{S}$ is a "projection" function with respect to some metric d'_E. In full generality, it is not guaranteed that there exists $\mathrm{P}_\mathfrak{S}$ such that $d'_E(\mathbf{x}^\star, \mathrm{P}_\mathfrak{S}(\mathbf{x}^\star)) = d'_E(\mathbf{x}^\star, \mathfrak{S})$, but one can always define it such that for all $\mathbf{x}^\star$, $d'_E(\mathbf{x}^\star, \mathrm{P}_\mathfrak{S}(\mathbf{x}^\star)) \leq d'_E(\mathbf{x}^\star, \mathfrak{S}) + \varepsilon$ for an arbitrary small $\varepsilon > 0$.*

Let us now introduce the proposed non-uniform IOP and LRIP.

Definition 3 (Non-uniform IOP). *A decoder Δ satisfies the non-uniform Instance Optimality Property for the (random) mapping Ψ, model $\mathfrak{S}$ and projection function $\mathrm{P}_\mathfrak{S}$, with constants $A, B > 0$, pseudometrics d_E, d'_E, probability $1-\rho$ and error $\lambda \geq 0$ if:*

$$\forall \mathbf{x}^\star \in E, \quad \mathbb{P}_\Psi\left[\forall \mathbf{e} \in F,\ d_E(\mathbf{x}^\star, \widetilde{\mathbf{x}}) \leq A d'_E(\mathbf{x}^\star, \mathrm{P}_\mathfrak{S}(\mathbf{x}^\star)) + B\|\mathbf{e}\|_F + \lambda\right] \geq 1 - \rho \tag{5}$$

where $\Delta(\Psi, \Psi(\mathbf{x}^\star) + \mathbf{e})$ is denoted by $\widetilde{\mathbf{x}}$.

[3] In this paper we assume that the minimization problem $\arg\min_{\mathbf{x}\in\mathfrak{S}} \|\Psi(\mathbf{x}) - \mathbf{y}\|_F$ has at least one solution, for simplicity (ties can be broken arbitrarily). When this is not the case, it is possible to consider a decoder that returns any element that approaches the infimum with a fixed precision, at the expense of having this precision in the decoding error λ, as in [8].

IOP Conf. Series: Journal of Physics: Conf. Series **1131** (2018) 012002 doi:10.1088/1742-6596/1131/1/012002

Note that in this definition the IOP is non-uniform with respect to the data $\mathbf{x}^\star$ but uniform with respect to the noise $\mathbf{e}$, meaning that with high probability on the draw of Ψ the (fixed) data can be stably recovered from a measurement vector with any additive noise.

Definition 4 (Non-uniform LRIP). *The operator Ψ satisfies the non-uniform LRIP for the model $\mathfrak{S}$ with constant $\alpha > 0$, pseudometric d_E, probability $1-\rho$ and error $\eta \geq 0$ if:*

$$\forall \mathbf{x} \in \mathfrak{S}, \quad \mathbb{P}_\Psi \Big[\forall \mathbf{x}' \in \mathfrak{S}, \ d_E(\mathbf{x}, \mathbf{x}') \leq \alpha \left\| \Psi(\mathbf{x}) - \Psi(\mathbf{x}') \right\|_F + \eta \Big] \geq 1 - \rho. \tag{6}$$

This LRIP is in fact "semi"-uniform: it is non-uniform with respect to one element $\mathbf{x}$ but uniform with respect to $\mathbf{x}'$. A "fully" non-uniform LRIP would, in fact, be almost always valid for many operators (see Section 4), and thus probably too weak to yield recovery guarantees.

Before stating our result, let us remark that the definition of the metric d'_E in Theorem 1 (*ii*) involves the operator Ψ, which is potentially problematic when it is random. To solve this, [13] introduces a so-called *Boundedness Property* (BP) in the classical sparse setting in finite dimension. We extend this notion in the considered context here.

Definition 5 (Boundedness property (BP)). *The operator Ψ satisfies the Boundedness Property with constant β, pseudometric d_G and probability $1-\rho$ if:*

$$\forall \mathbf{x} \in E, \ \forall \mathbf{x}_{\mathfrak{S}} \in \mathfrak{S}, \quad \mathbb{P}_\Psi \Big[\left\| \Psi(\mathbf{x}) - \Psi(\mathbf{x}_{\mathfrak{S}}) \right\|_F \leq \beta d_G(\mathbf{x}, \mathbf{x}_{\mathfrak{S}}) \Big] \geq 1 - \rho. \tag{7}$$

We then have the following result, proved in Appendix B.

Theorem 2 (The non-uniform LRIP and BP implies the non-uniform IOP). *Consider a random operator Ψ. Assume that:*

(i) the operator Ψ satisfies the non-uniform LRIP for the model $\mathfrak{S}$ with constant $\alpha > 0$, pseudometric d_E, probability $1-\rho_1$ and error $\eta \geq 0$;

(ii) the operator Ψ satisfies the non-uniform Boundedness Property with constant β, pseudometric d_G and probability $1-\rho_2$;

Then, the decoder Δ defined by (4) satisfies the non-uniform Instance Optimality Property for the operator Ψ, model $\mathfrak{S}$ and any projection function $\mathrm{P}_{\mathfrak{S}}$ with constants $A = 1$, $B = 2\alpha$, pseudometrics d_E and $d'_E := d_E + 2\alpha\beta d_G$, probability $1-\rho_1-\rho_2$ and error $\lambda = \eta$.

Compared with the result in [13], which proves non-uniform recovery under a *uniform* LRIP and the BP in the finite-dimensional case, our result holds under weaker hypotheses.

For the converse implication of Theorem 2, unlike Theorem 1, the non-uniform IOP does not seem to directly imply the non-uniform LRIP.

4. A typical proof of the LRIP

In this section, we outline a possible strategy to prove the LRIP, inspired by the proof for random matrices in [2]. This relatively simple proof has two steps: first, a pointwise concentration result, and second, an extension by covering arguments. For a set S and a metric d, we denote by $\mathcal{N}(S, d, \delta) \in \mathbb{N} \cup \{+\infty\}$ the minimum number of balls of radius δ, with centers that belong to S, required to cover S.

4.1. Linear case

We start with the linear case, which follows closely the proof in [2]. We treat the uniform case (Def. 2), with no error ($\eta = 0$). Assume (E, d_E) and $(F, \|\cdot\|_F)$ are both vector spaces, and that we have a random *linear* operator $\Psi : E \to F$ such that the following *concentration* result holds:

$$\forall \mathbf{x}, \mathbf{x}' \in \mathfrak{S} \text{ s.t. } d_E(\mathbf{x}, \mathbf{x}') > 0, \quad \mathbb{P}_\Psi \left(\tfrac{\|\Psi(\mathbf{x}) - \Psi(\mathbf{x}')\|_F}{d_E(\mathbf{x}, \mathbf{x}')} - 1 \leq -t \right) \leq e^{-c(t)} \tag{8}$$

for an increasing *concentration function* $c(t) > 0$. Typically, the "bigger" the space F is (*i.e.* the more measurements we collect), the higher the concentration function is: often, for m measurements ($F = \mathbb{R}^m$ or $\mathbb{C}^m$), classical concentration inequalities yield $c(t) \propto mt^2$.

This property proves a "pointwise" (or "fully" non-uniform) LRIP: for two given $\mathbf{x}, \mathbf{x}'$, the quantity $\|\Psi(\mathbf{x} - \mathbf{x}')\|_F$ is a good approximation of $d_E(\mathbf{x}, \mathbf{x})$ with high probability. We now invert the quantifiers by covering arguments. From the formulation of the concentration (8) we see that a particular set of interest is the so-called *normalized secant set* [23]:

$$\mathcal{S} = \left\{ \frac{\mathbf{x}-\mathbf{x}'}{d_E(\mathbf{x},\mathbf{x}')} \mid \mathbf{x}, \mathbf{x}' \in \mathfrak{S},\ d_E(\mathbf{x}, \mathbf{x}') > 0 \right\} \subset E \tag{9}$$

The proof of the following result is in Appendix C.

Proposition 1. *Consider* $0 < t < 1$. *Assume that the concentration property* (8) *holds, that* $\mathcal{S}$ *has finite covering numbers, and that for any draw of* Ψ *and any* $\mathbf{s}, \mathbf{s}' \in \mathcal{S}$ *we have* $\|\Psi(\mathbf{s} - \mathbf{s}')\|_F \leq C d_E(\mathbf{s}, \mathbf{s}')$. *Set* $\delta = \frac{t}{2C}$. *Define the probability of failure*

$$\rho = \mathcal{N}(\mathcal{S}, d_E, \delta) \cdot e^{-c(t/2)}$$

Then the operator Ψ *satisfies the uniform LRIP for the model* $\mathfrak{S}$ *with constant* $\alpha = (1-t)^{-1}$, *metric* d_E, *probability* $1 - \rho$ *and error* $\eta = 0$.

This proof of the RIP has been used for instance in classical compressive sensing [2] or for random linear embeddings of Radon measures [19]. It is also used in a constructive manner to build appropriate operators Ψ in [23].

4.2. Non-linear case

It is possible to adapt the previous proof to non-linear operators, by distinguishing the case where $\mathbf{x}$ and $\mathbf{x}'$ are "close", for which we resort to a linearization of Ψ and properties of the normalized secant set, and the case where $\mathbf{x}$ and $\mathbf{x}'$ are distant from each other, for which we use directly the covering numbers of the model. We treat here the non-uniform case (Def. 4).

Assume again that (E, d_E) and $(F, \|\cdot\|_F)$ are vector spaces. Assume that we have a random map $\Psi : E \to F$ such that the concentration property (8) holds. Next, suppose that there exists $\varepsilon_0 > 0$ such that for any fixed $\mathbf{x} \in \mathfrak{S}$:

(i) for all $\mathbf{x}' \in \mathfrak{S}$ and any draw of Ψ, $\|\Psi(\mathbf{x}) - \Psi(\mathbf{x}')\|_F \leq C_1 d_E(\mathbf{x}, \mathbf{x}')$

(ii) the model $\mathfrak{S}$ has finite covering numbers with respect to d_E, and in particular it also has finite diameter $M_{\mathfrak{S}} = \sup_{\mathbf{x},\mathbf{x}' \in \mathfrak{S}} d_E(\mathbf{x}, \mathbf{x}')$.

(iii) for all $0 < \varepsilon \leq \varepsilon_0$, the following version of the normalized secant set $\mathcal{S}_\varepsilon := \left\{ \frac{\mathbf{x}-\mathbf{x}'}{d_E(\mathbf{x},\mathbf{x}')} \mid \mathbf{x}' \in \mathfrak{S},\ 0 < d_E(\mathbf{x}, \mathbf{x}') \leq \varepsilon \right\}$ has finite covering numbers.

(iv) for all $\mathbf{x}' \in \mathfrak{S}$ such that $d_E(\mathbf{x}, \mathbf{x}') \leq \varepsilon_0$, and any draw of Ψ, we have $|\Psi(\mathbf{x}) - \Psi(\mathbf{x}') - D_\Psi(\mathbf{x} - \mathbf{x}')| \leq C_2 d_E(\mathbf{x}, \mathbf{x}')^2$, where $D_\Psi : E \to F$ is a linear map such that for all $\mathbf{s}, \mathbf{s}' \in \mathcal{S}_{\varepsilon_0}$, $\|D_\Psi(\mathbf{s} - \mathbf{s}')\|_F \leq C_3 d_E(\mathbf{s}, \mathbf{s}')$.

The following result is proved in Appendix D.

Proposition 2. *Assume that the properties* $(i) - (iv)$ *above are satisfied. Set* $\varepsilon \leq \min\left(\varepsilon_0, \frac{t}{8C_2}\right)$, $\delta' \leq \frac{t}{4C_3}$ *and* $\delta \leq \frac{t\varepsilon^2/(4C_1)}{\varepsilon + M_{\mathfrak{S}}}$. *Define the probability of failure*

$$\rho := \left[\mathcal{N}(\mathfrak{S}, d_E, \delta) + \mathcal{N}(\mathcal{S}_\varepsilon, d_E, \delta')\right] \cdot e^{-c(t/2)}$$

Then the operator Ψ *satisfies the non-uniform LRIP for the model* $\mathfrak{S}$ *with constant* $\alpha = (1-t)^{-1}$, *metric* d_E, *probability* $1 - \rho$ *and error* $\eta = 0$.

4.3. Illustration

In this section we illustrate the non-linear LRIP on a simple example; that of recovering a vector from a *random features embedding*, which is a random map initially designed for kernel approximation, see [24, 25]. Such a random embedding can be seen as a one-layer neural network with random weights, for which invertibility and preservation of information have recently been topics of interest [17, 18].

Consider $E = \mathbb{R}^d$ and define $\mathfrak{S}$ to be a *Union of Subspaces*, which is a popular model in compressed sensing [5], with controlled norm: $\mathfrak{S} = \bigcup_{i=1}^N (S_i \cap B_M)$ where $B_M = \{\mathbf{x},\ \|\mathbf{x}\|_2 \leq M\}$ and each S_i is an s-dimensional subspace of $\mathbb{R}^d$. As in [19], we choose a sampling that is a reweighted version of the original Fourier sampling for kernel approximation [24], for the Gaussian kernel with bandwidth $\sigma > 0$. It is defined as follows: for a number of measurements m, the measurements space is $F = \mathbb{C}^m$, the random map is defined as $\Psi(\mathbf{x}) = \frac{1}{\sqrt{m}}\left[\varphi_{\boldsymbol{\omega}_j}(\mathbf{x})\right]_{j=1}^m$ where $\varphi_{\boldsymbol{\omega}}(\mathbf{x}) = \frac{e^{\mathrm{i}\boldsymbol{\omega}_j^\top \mathbf{x}}}{f(\boldsymbol{\omega})}$ and $\boldsymbol{\omega}_j \in \mathbb{R}^d$ are drawn *i.i.d.* from $\Lambda = f(\boldsymbol{\omega})^2 \mathcal{N}(0, \sigma^{-2}\mathbf{I})$ (where $\mathcal{N}$ is a Gaussian), with $f(\boldsymbol{\omega})^2 = \frac{1}{3}\sum_{\ell=0}^2 \frac{\|\boldsymbol{\omega}\|_2^{2\ell}}{\gamma_{2\ell}}$ and $\gamma_\ell = \mathbb{E}_{\boldsymbol{\omega}\sim\mathcal{N}(0,\sigma^{-2}\mathbf{I})} \|\boldsymbol{\omega}\|_2^\ell = \mathcal{O}\left(\frac{d^{\ell/2}}{\sigma^\ell}\right)$. One can verify that $\int d\Lambda(\boldsymbol{\omega}) = 1$, *i.e.* Λ is a valid probability distribution. The metric d_E is here the *kernel metric* associated to the Gaussian kernel with bandwidth σ:

$$d_E(\mathbf{x}, \mathbf{x}') := 2\left(1 - \exp\left(-\frac{\|\mathbf{x}-\mathbf{x}'\|_2^2}{2\sigma^2}\right)\right)$$

We have the following result, which proof is in Appendix E.

Proposition 3. *If the number of measurements is such that*

$$m \gtrsim t^{-2} \cdot \left(s \cdot \log\left(\frac{Md}{\sigma t}\right) + \log(N) + \log\left(\frac{1}{\rho}\right)\right)$$

Then the operator Ψ satisfies the non-uniform LRIP for the model $\mathfrak{S}$ with constant $\alpha = (1-t)^{-1}$, metric d_E, probability $1-\rho$ and error $\eta = 0$, as well as the BP (Def. 5) with constant $\beta = 1 + t$, metric d_E and probability $1 - \rho$.

Hence, using Theorem 2, we have shown that a reduced number of random features preserves all information when encoding signals that are (in this case) well-modeled by a union of subspaces, *with respect to the associated kernel metric.* This preliminary analysis may have consequences for classical random feature bounds in a learning context [1, 26].

5. Conclusion

In this paper we generalized a classical property, the equivalence between the existence of an instance-optimal decoder and the LRIP, to non-linear inverse problems with possible quantization error or limited algorithmic precision, and data that live in any pseudometric set. We also formulated a version of the result for non-uniform recovery, by introducing a non-uniform version of the LRIP. To further illustrate this principle, we provided a typical proof strategy for the LRIP that one might use in practice, and gave an example of non-linear LRIP on random features for kernel approximation.

Although relatively simple in their proofs, these results may have important consequences for a large class of linear or non-linear inverse problems, where one seeks stable and robust recovery. Naturally, once the LRIP guarantees (or disproves) the existence of an instance optimal decoder, an outstanding question is the existence of *efficient* algorithms that provide equivalent guarantees, as in classical compressive sensing [16] or some of its recent extensions [27].

IOP Conf. Series: Journal of Physics: Conf. Series **1131** (2018) 012002 doi:10.1088/1742-6596/1131/1/012002

References

[1] Francis Bach. On the Equivalence between Kernel Quadrature Rules and Random Feature Expansions. *Journal of Machine Learning Research*, pages 1–38, 2017.

[2] Richard Baraniuk, Mark Davenport, Ronald A Devore, and Michael Wakin. A simple proof of the restricted isometry property for random matrices. *Constructive Approximation*, 28(3):253–263, 2008.

[3] Amir Beck and Yonina C. Eldar. Sparsity Constrained Nonlinear Optimization: Optimality Conditions and Algorithms. *SIAM Journal on Optimization*, 23(3):1480–1509, 2013.

[4] Thomas Blumensath. Compressed sensing with nonlinear observations and related nonlinear optimization problems. *Information Theory, IEEE Transactions on*, pages 1–19, 2013.

[5] Thomas Blumensath and Mike E. Davies. Sampling theorems for signals from the union of finite-dimensional linear subspaces. *IEEE Transactions on Information Theory*, 55(4):1872–1882, 2009.

[6] Petros T Boufounos and Richard G Baraniuk. 1-Bit Compressive Sensing. In *Information Sciences and Systems*, pages 16–21, 2008.

[7] Petros T. Boufounos, Shantanu Rane, and Hassan Mansour. Representation and Coding of Signal Geometry. *Information and Inference: a Journal of the IMA*, 6(4):349–388, 2017.

[8] Anthony Bourrier, Mike E. Davies, Tomer Peleg, and Rémi Gribonval. Fundamental performance limits for ideal decoders in high-dimensional linear inverse problems. *IEEE Transactions on Information Theory*, 60(12):7928–7946, 2014.

[9] Emmanuel J Candès. The restricted isometry property and its implications for compressed sensing. *Comptes Rendus Mathematique*, 346(9-10):1–4, 2008.

[10] Emmanuel J. Candès and Yaniv Plan. A probabilistic and RIPless theory of compressed sensing. *IEEE Transactions on Information Theory*, 57(11):7235–7254, 2011.

[11] Emmanuel J. Candès, Justin K. Romberg, and Terrence Tao. Robust uncertainty principles: Exact signal reconstruction from highly incomplete frequency information. *IEEE Transactions on Information Theory*, 52(2):480–509, 2006.

[12] Emmanuel J. Candès and Terrence Tao. Decoding by linear programming. *IEEE Transactions on Information Theory*, 51(12):4203–4215, 2005.

[13] Albert Cohen, Wolfgang Dahmen, and Ronald A Devore. Compressed sensing and best k-term approximation. *Journal of the American mathematical Society*, 22(1):211–231, 2009.

[14] David L. Donoho. Compressed sensing. *IEEE Transactions on Information Theory*, 52(4):1289–1306, 2006.

[15] Heinz W Engl and Philipp Kügler. Nonlinear inverse problems: theoretical aspects and some industrial applications. In *Multidisciplinary Methods for Analysis Optimization and Control of Complex Systems*, volume 06, pages 3–47. 2005.

[16] Simon Foucart and Holger Rauhut. *A Mathematical Introduction to Compressive Sensing.* Applied and Numerical Harmonic Analysis. Springer New York, NY, 2013.

[17] Anna C. Gilbert, Yi Zhang, Kibok Lee, Yuting Zhang, and Honglak Lee. Towards understanding the invertibility of convolutional neural networks. *IJCAI International Joint Conference on Artificial Intelligence*, pages 1703–1710, 2017.

[18] Raja Giryes, Guillermo Sapiro, and Alex M Bronstein. Deep Neural Networks with Random Gaussian Weights : A Universal Classification Strategy ? *IEEE Transactions on Signal Processing*, 64(13):3444–3457, 2016.

[19] Rémi Gribonval, Gilles Blanchard, Nicolas Keriven, and Yann Traonmilin. Compressive Statistical Learning with Random Feature Moments. *arXiv:1706.07180*, 2017.

[20] Laurent Jacques. Small width, low distortions: quasi-isometric embeddings with quantized sub-Gaussian random projections. pages 1–26, 2015.

[21] Nicolas Keriven, Anthony Bourrier, Rémi Gribonval, and Patrick Pérèz. Sketching for Large-Scale Learning of Mixture Models. *Information and Inference: A Journal of the IMA*, pages 1–62, 2017.

[22] Tomer Peleg, Rémi Gribonval, and Mike E. Davies. Compressed Sensing and Best Approximation from Unions of Subspaces: Beyond Dictionaries. *European Signal Processing Conference (EUSIPCO)*, 2013.

[23] Gilles Puy, Mike E. Davies, and Rémi Gribonval. Linear embeddings of low-dimensional subsets of a Hilbert space to R m. In *European Signal Processing Conference (EUSIPCO)*, pages 469–473, 2015.

[24] Ali Rahimi and Benjamin Recht. Random Features for Large Scale Kernel Machines. *Advances in Neural Information Processing Systems (NIPS)*, 2007.

[25] Ali Rahimi and Benjamin Recht. Weighted sums of random kitchen sinks: Replacing minimization with randomization in learning. *Advances in Neural Information Processing Systems (NIPS)*, 2009.

[26] Alessandro Rudi and Lorenzo Rosasco. Generalization Properties of Learning with Random Features. In *Advances in Neural Information Processing System (NIPS)*, 2017.

[27] Yann Traonmilin and Rémi Gribonval. Stable recovery of low-dimensional cones in Hilbert spaces : One RIP to rule them all. *Applied and Computational Harmonic Analysis*, 2016.

Appendix A. Proof of Theorem 1

(i) Consider $\mathbf{x}, \mathbf{x}' \in \mathfrak{S}$. By triangular inequality we have

$$d_E(\mathbf{x}, \mathbf{x}') \leq d_E(\mathbf{x}, \Delta(\Psi, \Psi(\mathbf{x}'))) + d_E(\Delta(\Psi, \Psi(\mathbf{x}')), \mathbf{x}').$$

Then, by applying the Instance Optimality Property with noise $\mathbf{e} := \Psi(\mathbf{x}') - \Psi(\mathbf{x})$ we get $d_E(\mathbf{x}, \Delta(\Psi, \Psi(\mathbf{x}')) \leq B \|\Psi(\mathbf{x}') - \Psi(\mathbf{x})\|_F + \lambda$, and by applying again the Instance Optimality Property it holds that $d_E(\Delta(\Psi, \Psi(\mathbf{x}')), \mathbf{x}') \leq \lambda$, hence the result.

(ii) Consider any signal $\mathbf{x}^\star \in E$ and noise $\mathbf{e} \in F$, denote $\mathbf{y} = \Psi(\mathbf{x}^\star) + \mathbf{e}$ and $\widetilde{\mathbf{x}} = \Delta(\Psi, \mathbf{y})$. Let $\mathbf{x}_{\mathfrak{S}} \in \mathfrak{S}$ be any element of the model. We have:

$$\begin{aligned} d_E(\mathbf{x}^\star, \widetilde{\mathbf{x}}) &\leq d_E(\mathbf{x}^\star, \mathbf{x}_{\mathfrak{S}}) + d_E(\mathbf{x}_{\mathfrak{S}}, \widetilde{\mathbf{x}}) \\ &\overset{\text{LRIP}}{\leq} d_E(\mathbf{x}^\star, \mathbf{x}_{\mathfrak{S}}) + \alpha \|\Psi(\mathbf{x}_{\mathfrak{S}}) - \Psi(\widetilde{\mathbf{x}})\|_F + \eta \\ &\leq d_E(\mathbf{x}^\star, \mathbf{x}_{\mathfrak{S}}) + \alpha \|\Psi(\mathbf{x}_{\mathfrak{S}}) - \mathbf{y}\|_F + \alpha \|\mathbf{y} - \Psi(\widetilde{\mathbf{x}})\|_F + \eta\,. \end{aligned}$$

By definition of the decoder (4) we have $\|\Psi(\widetilde{\mathbf{x}}) - \mathbf{y}\|_F \leq \|\Psi(\mathbf{x}_{\mathfrak{S}}) - \mathbf{y}\|_F$ and therefore

$$\begin{aligned} d_E(\mathbf{x}^\star, \widetilde{\mathbf{x}}) &\leq d_E(\mathbf{x}^\star, \mathbf{x}_{\mathfrak{S}}) + 2\alpha \|\Psi(\mathbf{x}_{\mathfrak{S}}) - \mathbf{y}\|_F + \eta \\ &\leq d_E(\mathbf{x}^\star, \mathbf{x}_{\mathfrak{S}}) + 2\alpha \|\Psi(\mathbf{x}_{\mathfrak{S}}) - \Psi(\mathbf{x}^\star)\|_F + 2\alpha \|\Psi(\mathbf{x}^\star) - \mathbf{y}\|_F + \eta \\ &\leq d'_E(\mathbf{x}^\star, \mathbf{x}_{\mathfrak{S}}) + 2\alpha \|\mathbf{e}\|_F + \eta \end{aligned}$$

where $d'_E(\mathbf{x}, \mathbf{x}') = d_E(\mathbf{x}, \mathbf{x}') + 2\alpha \|\Psi(\mathbf{x}) - \Psi(\mathbf{x}')\|_F$. Since the result is valid for all $\mathbf{x}_{\mathfrak{S}} \in \mathfrak{S}$, we can take the infimum of $d'_E(\mathbf{x}, \mathbf{x}_{\mathfrak{S}})$ with respect to $\mathbf{x}_{\mathfrak{S}} \in \mathfrak{S}$.

Appendix B. Proof of Theorem 2

Let $\mathbf{x}^\star \in E$ be a fixed signal and denote by $\mathbf{x}_{\mathfrak{S}} = \mathrm{P}_{\mathfrak{S}}(\mathbf{x}^\star) \in \mathfrak{S}$.

Applying the non-uniform LRIP, with probability at least $1 - \rho_1$ on the draw of the operator Ψ we have

$$\forall \mathbf{x}' \in \mathfrak{S},\ d_E(\mathbf{x}_{\mathfrak{S}}, \mathbf{x}') \leq \alpha \left\|\Psi(\mathbf{x}_{\mathfrak{S}}) - \Psi(\mathbf{x}'_{\mathfrak{S}})\right\|_F + \eta\,. \tag{B.1}$$

In the same fashion, applying the non-uniform Boundedness Property, with probability at least $1 - \rho_2$ on the draw of the operator Ψ we have

$$\|\Psi(\mathbf{x}^\star) - \Psi(\mathbf{x}_{\mathfrak{S}})\|_F \leq \beta d_G(\mathbf{x}^\star, \mathbf{x}_{\mathfrak{S}})\,. \tag{B.2}$$

Therefore, by a union bound, with probability at least $1 - \rho_1 - \rho_2$, both (B.1) and (B.2) are satisfied. If this is the case, for all noise vector $\mathbf{e} \in F$, denoting by $\mathbf{y} = \Psi(\mathbf{x}^\star) + \mathbf{e}$ and $\widetilde{\mathbf{x}} = \Delta(\Psi, \mathbf{y})$, it holds that:

$$\begin{aligned} d_E(\mathbf{x}^\star, \widetilde{\mathbf{x}}) &\leq d_E(\mathbf{x}^\star, \mathbf{x}_{\mathfrak{S}}) + d_E(\mathbf{x}_{\mathfrak{S}}, \widetilde{\mathbf{x}}) \\ &\overset{\text{(B.1)}}{\leq} d_E(\mathbf{x}^\star, \mathbf{x}_{\mathfrak{S}}) + \alpha \|\Psi(\mathbf{x}_{\mathfrak{S}}) - \Psi(\widetilde{\mathbf{x}})\|_F + \eta \\ &\leq d_E(\mathbf{x}^\star, \mathbf{x}_{\mathfrak{S}}) + \alpha \|\Psi(\mathbf{x}_{\mathfrak{S}}) - \mathbf{y}\|_F + \alpha \|\mathbf{y} - \Psi(\widetilde{\mathbf{x}})\|_F + \eta\,. \end{aligned}$$

Once again by definition of the decoder (4) we have $\|\Psi(\widetilde{\mathbf{x}}) - \mathbf{y}\|_F \leq \|\Psi(\mathbf{x}_{\mathfrak{S}}) - \mathbf{y}\|_F$ and therefore

$$\begin{aligned} d_E(\mathbf{x}^\star, \widetilde{\mathbf{x}}) &\leq d_E(\mathbf{x}^\star, \mathbf{x}_{\mathfrak{S}}) + 2\alpha \|\Psi(\mathbf{x}_{\mathfrak{S}}) - \mathbf{y}\|_F + \eta \\ &\leq d_E(\mathbf{x}^\star, \mathbf{x}_{\mathfrak{S}}) + 2\alpha \|\Psi(\mathbf{x}_{\mathfrak{S}}) - \Psi(\mathbf{x}^\star)\|_F + 2\alpha \|\Psi(\mathbf{x}^\star) - \mathbf{y}\|_F + \eta \\ &\leq d'_E(\mathbf{x}^\star, \mathbf{x}_{\mathfrak{S}}) + 2\alpha \|\mathbf{e}\|_F + \eta. \end{aligned}$$

by applying (B.2), which is the desired result.

Appendix C. Proof of Proposition 1

From the definition of the normalized secant set, our goal is to prove that with high probability on the draw of Ψ, for all $\mathbf{s} \in \mathcal{S}$ we have $\|\Psi \mathbf{s}\|_F \geq 1 - t$.

Let $\delta, \varepsilon > 0$ be small constants which values we shall define later and $N := \mathcal{N}(\mathcal{S}, d_E, \delta)$ be the covering numbers of the normalized secant set. Let $\mathbf{s}_1, \ldots, \mathbf{s}_N$ be an δ-covering of the normalized secant set $\mathcal{S}$.

By the concentration property, it holds that, with probability at least $1 - Ne^{-c(t/2)}$, for all $\mathbf{s}_i$ we have

$$\|\Psi \mathbf{s}_i\|_F \geq 1 - t/2 \tag{C.1}$$

Now, given any element of the normalized secant set $\mathbf{s} \in \mathcal{S}$, one can find an element of the covering $\mathbf{s}_i$ such that $\|\mathbf{s} - \mathbf{s}_i\|_E \leq \delta$. Assuming (C.1) holds, we have

$$\|\Psi \mathbf{s}\|_F \geq \|\Psi \mathbf{s_i}\|_F - \|\Psi (\mathbf{s} - \mathbf{s_i})\|_F \geq 1 - \frac{t}{2} - C\delta$$

and therefore by choosing $\delta = t/(2C)$ we obtain the desired result.

Appendix D. Proof of Proposition 2

Fix any $\mathbf{x}_\mathfrak{S} \in \mathfrak{S}$. Let $\delta, \delta', \varepsilon > 0$ be small constants which values we shall define later, such that $\varepsilon \leq \varepsilon_0$. Define $\mathfrak{S}_\varepsilon = \mathfrak{S} \backslash \{\mathbf{x} \in \mathfrak{S},\ d_E(\mathbf{x}_\mathfrak{S}, \mathbf{x}) \leq \varepsilon\}$ the model with a ball around $\mathbf{x}_\mathfrak{S}$ removed, and $N := \mathcal{N}(\mathfrak{S}_\varepsilon, d_E, \delta)$, $N' := \mathcal{N}(\mathcal{S}_\varepsilon, d_E, \delta')$. Let $\mathbf{x}_1, \ldots, \mathbf{x}_N$ and $\mathbf{s}_1, \ldots, \mathbf{s}_{N'}$ be, respectively, a δ-covering of $\mathfrak{S}_\varepsilon$ and a δ'-covering of $\mathcal{S}_\varepsilon$, with $\mathbf{x}'_i \in \mathfrak{S}$ defined such that $\mathbf{s}_i = \frac{\mathbf{x}_\mathfrak{S} - \mathbf{x}'_i}{d_E(\mathbf{x}_\mathfrak{S}, \mathbf{x}'_i)}$ and $0 < d_E(\mathbf{x}_\mathfrak{S}, \mathbf{x}'_i) \leq \varepsilon$.

By the concentration property, it holds that, with probability at least $1 - (N + N')e^{-c(t/2)}$, for all $\tilde{\mathbf{x}} = \mathbf{x}_i$ or $\tilde{\mathbf{x}} = \mathbf{x}'_i$ for all indices i, we have

$$\|\Psi(\mathbf{x}_\mathfrak{S}) - \Psi(\tilde{\mathbf{x}})\|_F \geq (1 - t/2) \cdot d_E(\mathbf{x}_\mathfrak{S}, \tilde{\mathbf{x}}). \tag{D.1}$$

Our goal is to extend this property to any element $\tilde{\mathbf{x}} = \mathbf{x}$ in the model $\mathfrak{S}$.

Let $\mathbf{x} \in \mathfrak{S}$ be any element of the model. We distinguish two cases. If $\mathbf{x} \in \mathfrak{S}_\varepsilon$, *i.e.* $d_E(\mathbf{x}_\mathfrak{S}, \mathbf{x}) \geq \varepsilon$, we consider $\mathbf{x}_i \in \mathfrak{S}_\varepsilon$ an element of the covering of $\mathfrak{S}_\varepsilon$ such that $d_E(\mathbf{x}, \mathbf{x}_i) \leq \delta$. We have:

$$\begin{aligned}
\left\| \tfrac{\Psi(\mathbf{x}_\mathfrak{S}) - \Psi(\mathbf{x})}{d_E(\mathbf{x}_\mathfrak{S}, \mathbf{x})} - \tfrac{\Psi(\mathbf{x}_\mathfrak{S}) - \Psi(\mathbf{x}_i)}{d_E(\mathbf{x}_\mathfrak{S}, \mathbf{x}_i)} \right\|_F &\leq \left\| \tfrac{\Psi(\mathbf{x}_\mathfrak{S}) - \Psi(\mathbf{x})}{d_E(\mathbf{x}_\mathfrak{S}, \mathbf{x})} - \tfrac{\Psi(\mathbf{x}_\mathfrak{S}) - \Psi(\mathbf{x}_i)}{d_E(\mathbf{x}_\mathfrak{S}, \mathbf{x})} \right\|_F + \left\| \tfrac{\Psi(\mathbf{x}_\mathfrak{S}) - \Psi(\mathbf{x}_i)}{d_E(\mathbf{x}_\mathfrak{S}, \mathbf{x})} - \tfrac{\Psi(\mathbf{x}_\mathfrak{S}) - \Psi(\mathbf{x}_i)}{d_E(\mathbf{x}_\mathfrak{S}, \mathbf{x}_i)} \right\|_F \\
&= \tfrac{\|\Psi(\mathbf{x}) - \Psi(\mathbf{x}_i)\|_F}{d_E(\mathbf{x}_\mathfrak{S}, \mathbf{x})} + \|\Psi(\mathbf{x}_\mathfrak{S}) - \Psi(\mathbf{x}_i)\|_F \left| \tfrac{1}{d_E(\mathbf{x}_\mathfrak{S}, \mathbf{x})} - \tfrac{1}{d_E(\mathbf{x}_\mathfrak{S}, \mathbf{x}_i)} \right| \\
&\overset{(i),(ii)}{\leq} \tfrac{C_1 \delta}{\varepsilon} + C_1 M_\mathfrak{S} \tfrac{d_E(\mathbf{x}, \mathbf{x}_i)}{d_E(\mathbf{x}_\mathfrak{S}, \mathbf{x}) d_E(\mathbf{x}_\mathfrak{S}, \mathbf{x}_i)} \\
&\leq \tfrac{C_1 \delta}{\varepsilon} \left(1 + \tfrac{M_\mathfrak{S}}{\varepsilon}\right)
\end{aligned}$$

and therefore

$$\left\| \frac{\Psi(\mathbf{x}_\mathfrak{S}) - \Psi(\mathbf{x})}{d_E(\mathbf{x}_\mathfrak{S}, \mathbf{x})} \right\|_F \geq 1 - t/2 - \frac{C_1 \delta}{\varepsilon} \left(1 + \frac{M_\mathfrak{S}}{\varepsilon}\right) \tag{D.2}$$

Now, when $d_E(\mathbf{x}_\mathfrak{S}, \mathbf{x}) \leq \varepsilon$, we define $\mathbf{z} := \frac{\mathbf{x}_\mathfrak{S} - \mathbf{x}}{d_E(\mathbf{x}_\mathfrak{S}, \mathbf{x})}$ and note that $\mathbf{z} \in \mathcal{S}_\varepsilon$. We approximate it by an element of the covering of the normalized secant set $\mathbf{s}_i = \frac{\mathbf{x}_\mathfrak{S} - \mathbf{x}'_i}{d_E(\mathbf{x}_\mathfrak{S}, \mathbf{x}'_i)} \in \mathcal{S}_\varepsilon$ (meaning that $d_E(\mathbf{x}_\mathfrak{S}, \mathbf{x}'_i) \leq \varepsilon$) that verifies $d_E(\mathbf{z}, \mathbf{s}_i) \leq \delta'$. Then, we have

$$\begin{aligned}
\left\| \tfrac{\Psi(\mathbf{x}_\mathfrak{S}) - \Psi(\mathbf{x})}{d_E(\mathbf{x}_\mathfrak{S}, \mathbf{x})} - \tfrac{\Psi(\mathbf{x}_\mathfrak{S}) - \Psi(\mathbf{x}'_i)}{d_E(\mathbf{x}_\mathfrak{S}, \mathbf{x}'_i)} \right\|_F &\leq \left\| \tfrac{\Psi(\mathbf{x}_\mathfrak{S}) - \Psi(\mathbf{x})}{d_E(\mathbf{x}_\mathfrak{S}, \mathbf{x})} - D_\Psi \mathbf{z} \right\|_F + \|D_\Psi(\mathbf{z} - \mathbf{s}_i)\|_F + \left\| D_\Psi \mathbf{s}_i - \tfrac{\Psi(\mathbf{x}_\mathfrak{S}) - \Psi(\mathbf{x}_i)}{d_E(\mathbf{x}_\mathfrak{S}, \mathbf{x}_i)} \right\|_F \\
&\overset{(iv)}{\leq} C_2 d_E(\mathbf{x}_\mathfrak{S}, \mathbf{x}) + C_3 \delta' + C_2 d_E(\mathbf{x}_\mathfrak{S}, \mathbf{x}'_i) \leq 2C_2 \varepsilon + C_3 \delta'
\end{aligned}$$

and therefore

$$\left\| \frac{\Psi(\mathbf{x}_\mathfrak{S}) - \Psi(\mathbf{x})}{d_E(\mathbf{x}_\mathfrak{S}, \mathbf{x})} \right\|_F \geq 1 - t/2 - (2C_2\varepsilon + C_3\delta') \tag{D.3}$$

To conclude, we set $\varepsilon \leq \min\left(\varepsilon_0, \frac{t}{8C_2}\right)$, $\delta' \leq \frac{t}{4C_3}$, $\delta \leq \frac{t\varepsilon^2/(4C_1)}{\varepsilon + M_\mathfrak{S}}$ to obtain the desired result.

Appendix E. Proof of Proposition 3

Concentration property. The concentration result is based on the fact that $\|\Psi(\mathbf{x}) - \Psi(\mathbf{x})\|_2^2 \approx \mathbb{E}_\omega |\varphi_\omega(\mathbf{x}) - \varphi_\omega(\mathbf{x}')|^2 = d_E(\mathbf{x}, \mathbf{x}')^2$ by definition of the random features. Using simple function studies and a Bernstein concentration inequality with a control on moments of all orders, it is possible to show (see [19], eq. (160) then Prop. 6.11) that the concentration result (8) is valid with $c(t) \propto \frac{mt^2}{1+t}$ (we do not reproduce the detailed proof here for brevity).

Since this Berstein inequality is in fact valid for all vectors (not necessarily in the model), as a consequence the Boundedness Property is also satisfied with constant $\beta = 1 + t$, metric d_E and probability $1 - e^{-c(t)}$.

We now check hypotheses (i)–(iv) in Proposition 2. We are going to repeatedly use the fact that for $\mathbf{x}, \mathbf{x}' \in \mathfrak{S}$, since $\|\mathbf{x}\|_2 \leq M$, we have

$$\ell \left\|\mathbf{x} - \mathbf{x}'\right\|_2 \leq d_E(\mathbf{x}, \mathbf{x}') \leq L \left\|\mathbf{x} - \mathbf{x}'\right\|_2 \tag{E.1}$$

where $\ell := 2 \cdot \frac{1 - \exp\left(-M^2/(2\sigma^2)\right)}{M}$ and $L = \frac{M}{\sigma^2}$.

(i) Using a first-order Taylor expansion we have $\|\Psi(\mathbf{x}) - \Psi(\mathbf{x}')\|_2 \leq \left(\sum_{j=1}^m \frac{\|\boldsymbol{\omega}_j\|_2^2}{mf(\boldsymbol{\omega}_j)^2}\right)^{\frac{1}{2}} \|\mathbf{x} - \mathbf{x}'\|_2 \lesssim \sqrt{\gamma_2}\, \|\mathbf{x} - \mathbf{x}'\|_2$ hence by (E.1) hypothesis (i) is valid with $C_1 \propto \frac{\sqrt{\gamma_2}}{\ell}$.

(ii) It is immediate that $M_\mathfrak{S} = ML$. Then, using the well-known fact that for each S_j we have $\mathcal{N}(S_j \cap B_M, \|\cdot\|_2, \delta) = \mathcal{O}\left(\left(\frac{M}{\delta}\right)^s\right)$, by a union bound and (E.1) we have $\mathcal{N}(\mathfrak{S}, d_E, \delta) \leq \mathcal{O}\left(N\left(\frac{LM}{\delta}\right)^s\right)$.

(iii) Similar to the model, for all ε the normalized secant set is included in a union of subspace $\mathcal{S}_\varepsilon \subset \bigcup_{i,j=1}^N S_{ij} \cap B_{M/\ell}$, with norm controlled by M/ℓ by (E.1), and where $S_{ij} = S_i + S_j$ (sum of subspaces). Hence, since $\mathcal{N}(S_{ij} \cap B_{M/\ell}, \|\cdot\|_2, \delta) = \mathcal{O}\left(\left(\frac{M}{\ell\delta}\right)^{2s}\right)$, by a union bound and (E.1) we have $\mathcal{N}(\mathcal{S}_\varepsilon, d_E, \delta) \leq \mathcal{O}\left(N^2\left(\frac{LM}{\ell\delta}\right)^{2s}\right)$.

(iv) Finally, by a Taylor expansion we have

$$\varphi_\omega(\mathbf{x}) - \varphi_\omega(\mathbf{x}') = \frac{\mathrm{i}\boldsymbol{\omega}^\top}{f(\boldsymbol{\omega})}(\mathbf{x} - \mathbf{x}')e^{\mathrm{i}\boldsymbol{\omega}^\top\mathbf{x}} - \frac{1}{2} \cdot \frac{(\boldsymbol{\omega}^\top(\mathbf{x} - \mathbf{x}'))^2}{f(\boldsymbol{\omega})}e^{\mathrm{i}\boldsymbol{\omega}^\top\mathbf{x}''}$$

where $\mathbf{x}'' = \theta\mathbf{x} + (1 - \theta)\mathbf{x}'$. Hence hypothesis (iv) is satisfied with $C_2 = C_1$ and $C_3 = \frac{\sqrt{\gamma_4}}{\ell^2}$, which concludes the proof.

IOP Conf. Series: Journal of Physics: Conf. Series **1131** (2018) 012003 doi:10.1088/1742-6596/1131/1/012003

Human chest imaging by real-time processing of electrical impedance data tomography

G Gottardi[1] **and L Poli**[1]

[1] ELEDIA Research Center (ELEDIA@UniTN - University of Trento)
Via Sommarive 9, I-38123 Trento, Italy

E-mail: lorenzo.poli@unitn.it

Abstract. Real-time human chest imaging exploiting electrical impedance tomography (*EIT*) data is addressed in this work. Robust estimations of the lungs conductivity, directly related to their air/liquid content, are obtained by formulating the arising inverse problem within the learning-by-examples (*LBE*) framework. The partial least squares (*PLS*) algorithm is exploited to reduce the dimensionality of the feature space, while an adaptive sampling strategy is exploited to build an optimal training set of input/output pairs used to build a computationally efficient surrogate model of the inverse operator. Selected numerical results are shown to assess the effectiveness and the potentialities of the proposed *LBE* strategy.

1. Introduction

Electrical impedance tomography (*EIT*) is a promising alternative to computed tomography (*CT*) and X-ray radiography in many medical non-invasive diagnosis applications including breast cancer detection and heart/brain/lungs activity monitoring [1]-[6]. Differently from both standard investigation technologies, *EIT* provides lower resolution images of the investigated domain but it doesn't expose the patient to ionizing radiation. Moreover, it allows a continuous low-cost and easy-deployable monitoring, which is fundamental in some clinical scenarios such as, for example, the monitoring of the air content of the lungs in patients under mechanical ventilation in intensive-care units. Within this applicative context, low-frequency currents are induced in the chest through pairs of electrodes attached to the patient skin. The conductivity distribution within the inspected domain (strongly related to the air/liquid content of the lungs) determines the resulting voltages measured at other pairs of electrodes, allowing to retrieve - through suitable processing - a cross-sectional image of the human thorax [1]-[6].

The poor signal-to-noise ratio as well as the limited number of independent measurements lead to a severely *non-linear* and *ill-posed* inverse problem, for which suitable reconstruction algorithms and regularization schemes have to be carefully developed and/or customized. As an example, Tikhonov-based [5] or sparseness-based [6]-[10] regularizations can be successfully exploited to address the solution of the inverse *EIT* problem. Alternatively, learning-by-examples (*LBE*) methods can be regarded as computationally-efficient approaches to achieve real-time guesses of the imaged domain without exploiting iterative (costly) evaluations of the forward operator [11]-[19].

LBE strategies are based on a two-step procedure aimed at (*a*) generating a suitable set of input/output (*I/O*) examples in order to train a surrogate model during the *off-line phase* and (*b*) exploiting such a trained model to make fast and accurate predictions during the *on-line phase* starting from previously-unseen data [11]. Within this context, this work presents a novel *LBE* strategy for the real-time estimation the conductivity of the human lungs from *EIT* measurements. The developed inversion technique exploits the innovative integration of the partial least squares (*PLS*) feature extraction technique with an adaptive sampling scheme aimed at generating an optimal (i.e., highly-informative and reduced-size) training set [13], which is in turn used to train an augmented radial basis function (*A-RBF*) predictor [20].

2. Mathematical formulation

Let us consider a two-dimensional bounded investigation domain D representing a cross-section of the human thorax (Fig. 1). The conductivity distribution within D, $\sigma(x,y)$, is assumed to be purely real (resistive) and piecewise constant depending on the organ/tissue (i.e., lungs, muscle, fat, and hearth - Fig. 1).

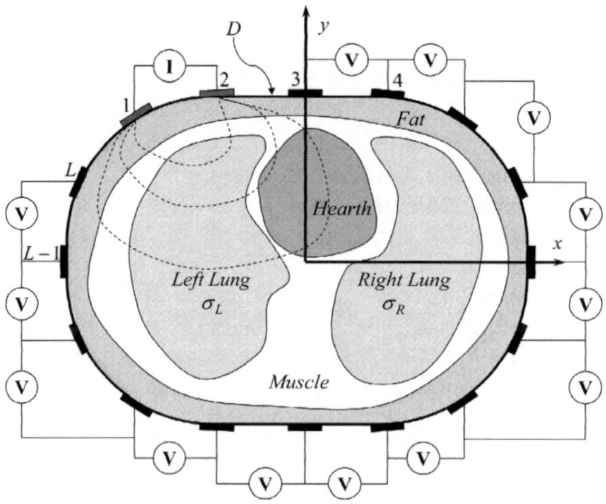

Figure 1. Geometry of the *2D* human thorax model and *EIT* acquisition system.

The adjacent electrode pair measurement strategy [2] is considered to collect *EIT* data exploiting a set of L electrodes uniformly displaced over the external boundary of D (Fig. 1). Accordingly, a low-frequency current (i.e., $f < 100\,\text{kHz}$) is successively applied to each pair of electrodes, the resulting voltages being measured over all remaining $(L-3)$ adjacent pairs of electrodes (Fig. 1). The total number of independent measurements is thus equal to [1]

$$M = \frac{L(L-3)}{2} \tag{1}$$

A standard finite element method (*FEM*) is used to numerically treat the forward problem, by sub-dividing D into N triangular mesh cells [2]. Accordingly, the set of measured voltages corresponding to a given conductivity distribution within D, $\boldsymbol{\sigma} = \{\sigma(x_n, y_n); n = 1,\ldots,N\}$, is computed exploiting the low-frequency approximation through the Poisson equation [2], resulting in

$$\mathbf{V} = \Phi\{\boldsymbol{\sigma}\} \tag{2}$$

where $\mathbf{V} = \{V_m; m = 1,\ldots,M\}$ and $\Phi\{.\}$ represents the forward *2D EIT-FEM* operator.

The goal of the inverse problem is to estimate the conductivity value inside the regions occupied by the left ($\sigma(x,y) = \sigma_L$) and right ($\sigma(x,y) = \sigma_R$) lungs, which is directly proportional to the their air/liquid content, by assuming that the conductivity of the neighboring regions (i.e., fat, muscle, and heart - Fig. 1) is fixed and a-priori known. In other words, a suitable inversion scheme must be adopted in order to estimate the unknown vector $\mathbf{\Omega} = (\sigma_L, \sigma_R)$ starting from the corresponding set of *EIT* voltages, i.e., solving the following implicit-form expression

$$\widetilde{\mathbf{\Omega}} = \widetilde{\Phi}^{-1}\{\mathbf{V}\} \tag{3}$$

where the hat "~" indicates that inverse operator $\Phi^{-1}\{.\}$ can be only approximated since it is unknown. To solve the inverse problem (3) an innovative inversion strategy is proposed in the framework of the learning-by-examples (*LBE*) theory [11]. Thanks to such a technique, the information about the two lungs embedded inside *EIT* measurements can be effectively *compressed* into a very small-dimensional features space, which is in turn adaptively sampled by an iterative sampling scheme in order to accurately model the *I/O* relationship with a limited set of training samples/observations. The solution of (3) is obtained through the following procedural steps.

1. *Initialization*. In order to adaptively sampling the two-dimensional space identified by $\mathbf{\Omega}$, the first step is to generate an initial set of *I/O* pairs. Towards this end, the Latin hypercube sampling (*LHS*) strategy is applied to randomly sample the input space and generate $S = S_0$ configurations (within predefined variation ranges)

$$\mathbf{\Omega}^s = \left(\sigma_L^s, \sigma_R^s\right), \; s = 1, \ldots, S_0 . \tag{4}$$

For each input sample, the forward *EIT-FEM* problem is solved in order to compute the corresponding *EIT* data, i.e.,

$$\mathbf{V}^s = \left(V_m^s; m = 1, \ldots, M\right) = \Phi\left\{\mathbf{\Omega}^s\right\}, \; s = 1, \ldots, S_0 . \tag{5}$$

Both input samples and corresponding outputs are then used to fill two matrices, i.e.,

$$\boldsymbol{\sigma}_0 = \begin{bmatrix} \sigma_L^1 & \sigma_R^1 \\ \sigma_L^2 & \sigma_R^2 \\ \vdots & \vdots \\ \sigma_L^{S_0} & \sigma_R^{S_0} \end{bmatrix}; \quad \mathbf{V}_0 = \begin{bmatrix} V_1^1 & V_2^1 & \cdots & V_M^1 \\ V_1^2 & \ddots & & \vdots \\ \vdots & & & \vdots \\ V_1^{S_0} & V_2^{S_0} & \cdots & V_M^{S_0} \end{bmatrix} . \tag{6}$$

2. *Feature Extraction*. Instead of directly exploiting the *M*-dimensional measured voltages for solving the inverse problem, the information about $\boldsymbol{\sigma}_0$ embedded within $\mathbf{V}_0$ is *compressed* into a small set of $J << M$ features through the partial least squares (*PLS*) algorithm. Thanks to such an approach, it is possible to (*i*) mitigate the *curse of dimensionality* [11] when solving the inverse problem, since a lower number of training samples will be necessary for an accurate modeling of $\Phi^{-1}\{.\}$, and (*ii*) filter a large amount of noise normally affecting *EIT* measurements. Accordingly, the SIMPLS algorithm [13] is exploited to compute from $\boldsymbol{\sigma}_0$ and $\mathbf{V}_0$ the *PLS* weight matrix $\mathbf{W}$ allowing to compute the *extracted features matrix*

$$\mathbf{T}_0 = \begin{bmatrix} T_1^1 & T_2^1 & \cdots & T_J^1 \\ T_1^2 & \ddots & & \vdots \\ \vdots & & & \vdots \\ T_1^{S_0} & T_2^{S_0} & \cdots & T_J^{S_0} \end{bmatrix} \tag{7}$$

through a linear combination of the original *measured features*, i.e.,

$$\mathbf{T}_0 = \mathbf{V}_0 \mathbf{W} . \tag{8}$$

Each row of $\mathbf{T}_0$ contains the *J* extracted features associated to a given input (σ_L, σ_R) configuration, and can be used in substitution to the corresponding (*M*-dimensional) row of $\mathbf{V}_0$. The initial training set is then formed with the generated S_0 *I/O* pairs as follows

$$\mathbf{\Psi}_0 = \left\{\left(\left(T_j^s; j = 1, \ldots, J\right), \left(\sigma_L^s, \sigma_R^s\right)\right); s = 1, \ldots, S_0\right\} \tag{9}$$

IOP Conf. Series: Journal of Physics: Conf. Series **1131** (2018) 012003 doi:10.1088/1742-6596/1131/1/012003

3. *Adaptive Sampling*. Once the *PLS* matrix $\mathbf{W}$ has been determined, it is possible to map additional *EIT* measurements into the *J*-dimensional space of extracted features. The generation of new training samples is done by adaptively sampling the $\mathbf{\Omega}$ space, such that a uniform coverage of the extracted features space is obtained [13]. At each iteration of such an adaptive scheme, a set of C candidates, $\left(\sigma_L^c,\sigma_R^c\right)$, $c=1,...,C$, is generated via *LHS* sampling. The goal is to select the candidate $\mathbf{\Omega}^*$ that exhibits the largest minimum Euclidean distance in the extracted features space to all previously collected *S* training samples (being $S=S_0$ at the first iteration), i.e.,

$$\mathbf{\Omega}^*=\left(\sigma_L^*,\sigma_R^*\right)=\arg\left(\max_{c=1,...,C}\left\{\min_{s=1,...,S}\sqrt{\sum_{j=1}^{J}\left(T_j^c-T_j^s\right)^2}\right\}\right). \quad (10)$$

In order to avoid the evaluation of the extracted features associated to each candidate, which would require the repeated solution of the forward *EIT-FEM* problem (with a significant waste of computational resources), a fast linear interpolator is exploited on the existing training samples to predict T_j^c, $j=1,....,J$, $c=1,...,C$ [13]. Then, once $\mathbf{\Omega}^*$ has been identified through (10), the actual *PLS* features associated to such sample are determined by computing (now only once) the associated *EIT* voltages, i.e., $\mathbf{V}^*=\left(V_m^*;m=1,...,M\right)=\Phi\left\{\sigma_L^*,\sigma_R^*\right\}$, and transforming them through the matrix $\mathbf{W}$ (i.e., $\mathbf{T}^*=\left(T_j^*;j=1,...,J\right)=\mathbf{V}^*\mathbf{W}$). The training set is then updated by adding the selected *I/O* pair

$$\mathbf{\Psi}_S=\mathbf{\Psi}_{S-1}\cup\left\{\mathbf{T}^*;\mathbf{\Omega}^*\right\} \quad (11)$$

and the process is iterated until a predefined stopping criterion is met (e.g., when a maximum number of forward solver evaluations, S_{max}, is reached).

4. *Prediction*. Once the training set has been adaptively generated, it can be used to build a prediction model which will be able to estimate the lungs conductivities starting from previously-unseen *EIT* measurements. In this work, the augmented radial basis function (*A-RBF*) technique [20] is used to build such a model given its numerical efficiency and the need to calibrate only a single parameter, i.e., the width of the *RBF* kernel function, γ. During the on-line test phase, each new test *EIT* measurement is first converted through $\mathbf{W}$ to the corresponding set of *J* extracted features, and then given as input to the trained *A-RBF* model to estimate the corresponding lungs conductivities.

3. Numerical validation

In order to verify the effectiveness of the proposed *LBE* inversion strategy for real-time *EIT* lungs monitoring, a set of numerical experiments is carried out considering $L=16$ electrodes surrounding *D*, resulting into $M=104$ independent voltage measurements. The human thorax is sub-divided into $N=3130$ triangular mesh cells, while the conductivity of the heart, fat, and muscle are set according to the low−frequency tissue properties table of the IT'IS Foundation [21] to 3.81×10^{-1} S/m, 5.73×10^{-2} S/m, and 3.55×10^{-1} S/m, respectively (Fig. 1). To build the training set, the left and right lungs conductivities are varied within the range $\left(\sigma_L,\sigma_R\right)\in\left[4.18\times10^{-2},8.05\times10^{-1}\right]$ S/m. The initial training set consists of $S_0=25$ *LHS*-generated samples, while $J=2$ is the number of *PLS*-extracted features (carefully set after a preliminary calibration process). To test the robustness of the inversions against noise, an additive white Gaussian noise is considered to corrupt *EIT* measurements with different signal-to-noise-ratios (*SNR*s).

To provide a pictorial view of the achievable estimation accuracy, Figure 2 shows the scatter plot of the actual versus predicted conductivity for the left [Figs. 2(*a*),2(*c*)] and right [Figs. 2(*b*),2(*d*)] lungs when training the *A-RBF* with $S=S_{max}=225$ training samples and testing it over $K=1000$ *LHS*-

IOP Conf. Series: Journal of Physics: Conf. Series **1131** (2018) 012003 doi:10.1088/1742-6596/1131/1/012003

generated noisy *EIT* test measurements at $SNR = 40$ dB [Figs. 2(*a*)-2(*b*)] and $SNR = 20$ dB [Figs. 2(*c*)-2(*d*)].

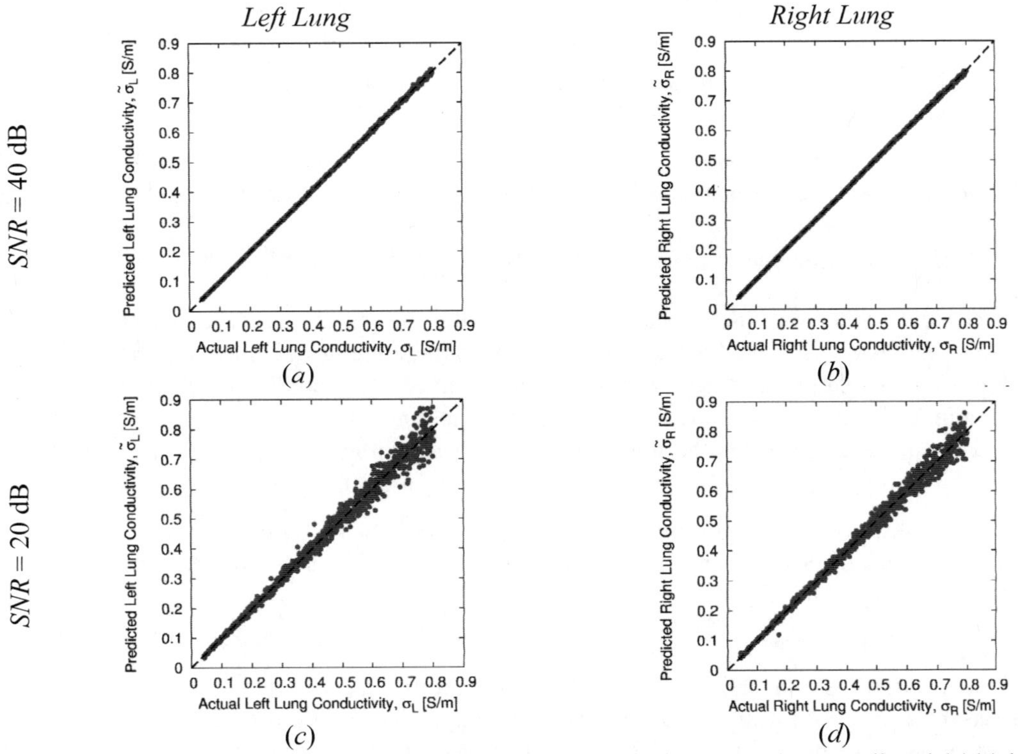

Figure 2. *Numerical Assessment* ($M = 104$, $J = 2$, $S = 225$) - Actual vs. predicted (*a*)(*c*) left and (*b*)(*d*) right lung conductivity when processing noisy data at (*a*)(*b*) $SNR = 40$ dB and (*c*)(*d*) $SNR = 20$ dB.

As it can be observed, a very high prediction accuracy is yielded at both levels of noise, with accurate guesses of the left and right conductivities within the considered variation ranges.

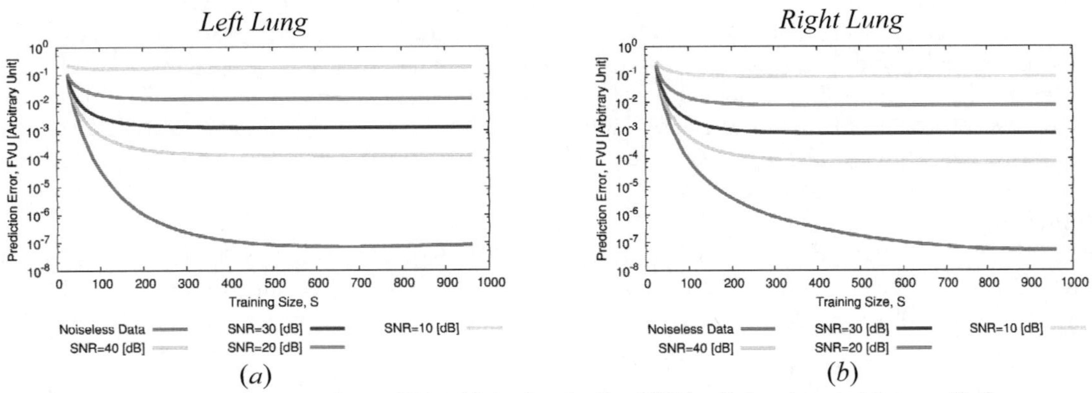

Figure 3. *Numerical Assessment* ($M = 104$, $J = 2$, $S = 225$) - Behaviour of the prediction error versus the number of training samples for (*a*) left and (*b*) left lung conductivity estimation.

To have a wider overview of the achievable prediction performances, Figure 3 shows the behaviour of the prediction error versus the number of training samples, *S*, used to train the *A-RBF*. The reported error is defined as the fraction of unexplained variance (*FVU*), computed as follows

$$FVU = \frac{\sum_{k=1}^{K}\left(\sigma_k - \widetilde{\sigma}_k\right)^2}{\sum_{k=1}^{K}\left(\sigma_k - \sigma_{av}\right)^2} \tag{12}$$

where σ_k and $\widetilde{\sigma}_k$ are the actual and predicted conductivity for the *k*-th test sample, and $\sigma_{av} = \frac{1}{K}\sum_{k=1}^{K}\sigma_k$ is the average conductivity within the test samples. It is worth observing that the error rapidly drops as new training samples are adaptively added to the initial ones (Fig. 3). Moreover, the error turns out to be always $FVU < 1$ whatever the *SNR* and the training size, meaning that the *A-RBF* is capable of providing better estimations with respect to the most naive predictor (which would output for each test sample the average conductivity value, thus yielding $FVU = 1$). As for what concerns the required computational times for the inversion, the *PLS* algorithm extracts the *J* features from a new measurement within fraction of seconds ($\Delta t^{PLS} \approx 10^{-2}$ sec), while the *A-RBF* is able to output a prediction of the lungs conductivities in about $\Delta t^{A-RBF} \approx 10^{-1}$ sec. Accordingly, it is possible to confirm that the whole inversion process is almost performed in real-time as soon as new *EIT* data is available from the acquisition system.

4. Conclusions

The computationally-efficient inversion of *EIT* data for monitoring the air content in human lungs has been addressed through an innovative *LBE* inversion strategy. Thanks to the *PLS* feature extraction a significant compression of the informative features can be obtained, allowing both a reduction of the necessary training samples to accurately model the *I/O* relation and an effective noise filtering during the on-line test phase. It should be remarked that the proposed methodology is not aimed at retrieving a full image of the conductivity distribution within the chest (as done in classical imaging), but rather at estimating the value of some monitored parameters of clinical relevance. Of course, these latter are not necessarily limited to the lungs conductivities, since the proposed approach is very flexible and customizable to the needs of the specific application. Selected numerical results have been presented to assess the effectiveness of the proposed strategy, as well as to confirm its capability of performing almost instantaneous inversions. Future works will be devoted at extending the capabilities of the *LBE* method to estimate other parameters of interest within the human thorax form *EIT* data, as well as to consider a fully-*3D* modeling of the forward problem. Moreover, further studies will be devoted at studying the robustness of the linearity assumption made by the *PLS* algorithm, as well as to validate the approach against real *EIT* data.

Acknowledgment
This work has been partially supported by the Italian Ministry of Foreign Affairs and International Cooperation, Directorate General for Cultural and Economic Promotion and Innovation within the SNATCH Project (2017-2019).

References

[1] Brown BH 2003 Electrical impedance tomography (EIT): a review *J. Med. Eng. Technol.* **27** 97-108
[2] Denaï MA, Mahfouf M, Mohamad-Samuri S, Panoutsos G, Brown BH and Mills GH 2010 Absolute electrical impedance tomography (aEIT) guided ventilation therapy in critical care

patients: simulations and future trends *IEEE Trans. Inf. Technol. Biomed.* **14** 641-649

[3] Borcea L 2002 Electrical impedance tomography *Inverse Probl.* **18** R99-R136

[4] Moura FS, Aya JCC, Fleury AT, Amato MBP and Lima RG 2010 Dynamic imaging in electrical impedance tomography of the human chest with online transition matrix identification *IEEE Trans. Biomed. Eng.* **57** 422-431

[5] Vauhkonen M, Vadasz D, Karjalainen PA, Somersalo E and Kaipio JP 1998 Tikhonov regularization and prior information in electrical impedance tomography *IEEE Trans. Med. Imag.* **17** 285-293

[6] Gehre M, Kluth T, Lipponen A, Jin B, Seppänen A, Kaipio J and Maass P 2012 Sparsity reconstruction in electrical impedance tomography: an experimental evaluation *J. Comput. Appl. Math.* **236** 2126-2136

[7] Oliveri G, Salucci M, Anselmi N and Massa A 2017 Compressive sensing as applied to inverse problems for imaging: theory, applications, current trends, and open challenges *IEEE Antennas Propag. Mag.* **59** 34-46

[8] Anselmi N, Oliveri G, Hannan M A, Salucci M and Massa A 2017 Color compressive sensing imaging of arbitrary-shaped scatterers *IEEE Trans. Microw. Theory Techn.* **65** 1986-1999

[9] Anselmi N, Oliveri G, Salucci M and Massa A 2015 Wavelet-based compressive imaging of sparse targets *IEEE Trans. Antennas Propag.* **63** 4889-4900

[10] Salucci M, Gelmini A, Poli L, Oliveri G and Massa A 2018 Progressive compressive sensing for exploiting frequency-diversity in GPR imaging *Journal of Electromagnetic Waves and Applications,* in press

[11] Massa A, Oliveri G, Salucci M, Anselmi N and Rocca P 2018 Learning-by-examples techniques as applied to electromagnetics *J. Electromagn. Waves Appl.* **32** 516-541

[12] Demidenko E, Hartov A and Paulsen K 2004 Statistical estimation of resistance/conductance by electrical impedance tomography measurements *IEEE Trans. Med. Imag.* **23** 829-838

[13] Salucci M, Anselmi N, Oliveri G, Calmon P, Miorelli R, Reboud C and Massa A 2016 Real-time NDT-NDE through an innovative adaptive partial least squares SVR inversion approach *IEEE Trans. Geosci. Remote Sens.* **54** 6818-6832

[14] Salucci M, Vrba J, Merunka I and Massa A 2017 Real-time brain stroke detection through a learning-by-examples technique - an experimental assessment *Microw. Opt. Technol. Lett.* **59** 2796-2799

[15] Li M, Abubakar A and Habashy TM 2012 A three-dimensional model-based inversion algorithm using radial basis functions for microwave data *IEEE Trans. Antennas Propag.* **60** 3361-3372

[16] Wu Y, Tang ZX, Zhang B and Xu Y 2007 Permeability measurement of ferromagnetic materials in microwave frequency range using support vector machine regression *Prog. Electromagn. Res.* **70** 247-256

[17] Ahmed S, Salucci M, Miorelli R, Anselmi N, Oliveri G, Calmon P, Reboud C and Massa A 2017 Real time groove characterization combining partial least squares and SVR strategies: application to eddy current testing *J. Phys. Conf. Ser.* **904** 1-8

[18] Ahmed S, Miorelli R, Salucci M and Massa A 2017 Real-time flaw characterization through learning-by-examples techniques: a comparative study applied to ECT *Electromagnetic Nondestructive Evaluation (XX)* **42** 228-235

[19] Viani F, Migliore MD, Polo A, Salucci M and Massa A 2018 Iterative classification strategy for multi-resolution wireless sensing of passive targets *Electron. Lett.* **54** 101-103

[20] Sui YK, Li SP and Guo YQ 2008 An efficient global optimization algorithm based on augmented radial basis function *Int. J. Simul. Multisci. Des. Optim.* **2** 49-55

[21] IT'IS Foundation, Low Frequency (Conductivity) Tissue Properties, https://www.itis.ethz.ch/virtual-population/tissue-properties/database/low-frequency-conductivity

IOP Conf. Series: Journal of Physics: Conf. Series **1131** (2018) 012004 doi:10.1088/1742-6596/1131/1/012004

CS-based computational methods for inverse problems arising in arrays processing and design

A Massa[1,2], G Gottardi[1] and E Rajo-Iglesias[2]

[1] ELEDIA Research Center (ELEDIA@UniTN - University of Trento)
Via Sommarive 9, I-38123 Trento, Italy
[2] ELEDIA Research Center (ELEDIA@UC3M - Universidad Carlos III de Madrid)
Avenida de la Universidad 30, 28911 Leganés, Madrid – Spain

E-mail: andrea.massa@unitn.it

Abstract. This work presents a review of the recent Compressive Sensing (*CS*)-based computational methodologies when applied to the design and processing of antenna arrays. The arising inverse problems have been properly formulated in order to deals with the *CS* theoretical requirements. Two instances of array processing and design, namely the diagnosis of failed elements and the synthesis of clustered arrays, are proposed and validated by a simple numerical example.

1. Introduction

In the last years the Compressive Sensing (*CS*) paradigm has been proposed as a promising and effective strategy for solving inverse problems in different research fields, from video/image processing to electromagnetics [1]-[3]. Starting from a reduced set of measurements, fewer than required from standard Nyquist based strategies, *CS* allows fast and reliable inversions, assuming the sparsity of the unknowns [1]. Anyway the use of *CS* in generic inverse problems is not trivial: the choice of the basis function used for the sparse representation of the unknowns is of non-negligible importance [3], and the problem matrix, linking the measurements to the unknowns, must satisfy the restrict isometry property (*RIP*) [3][4]. Unfortunately checking the *RIP* condition is not a simple task and in most of the cases it is computationally unfeasible. Indeed the problem must be properly formulated in order to comply with the *CS* theoretical requirements [2][3]. As instance, *CS* has been recently applied to free-space inverse scattering problems [5]-[8] by mans of Bayesian formulations, for the imaging of intrinsically sparse (i.e. pixel sparse) targets [5], and of scatterers that are sparse in transformed domains [6]-[8]. The effectiveness of *CS* methodologies in solving other electromagnetics related problems [2] has been effectively showed when applied to the synthesis of sparse arrays of antennas [9]-[17], considering linear [10], planar [11]-[14] and conformal geometries [16][17], and properly handling the complex nature of the unknown variables [10]. *CS* has been also successfully exploited for the estimation of direction-of-arrivals (*DoA*) signals [18]-[21], thanks to the intrinsic sparsity of the *DoA*s in the angular domain, considering single [18][19] and multiple snapshot techniques [20][21], and a multi-resolution scheme for real elements arrays affected by mutual-coupling [19].

In this work the recent advancements of *CS* as applied to inverse problems arising in array processing and synthesis are reported. More in detail, the diagnosis problem of failures in planar phased arrays [22]-[25] and the design of contiguously clustered linear arrays [26] are formulated and validated with a set of numerical examples. Finally the limitations and the future trends of *CS* applied to array processing related inverse problems are discussed as well.

2. Mathematical Formulation

Let us consider an array of P elements having barycenters $\mathbf{r}_p = (x_p, y_p)$, $p = 1,...,P$ weighted by a set of complex coefficients $\mathbf{w} = \{w_p; p = 1,...,P\}$ related to the corresponding radiated pattern as:

$$\mathbf{F} = \mathbf{A}\mathbf{w} \tag{1}$$

where in (1) $\mathbf{A} = \left\{a_{np} = e^{jk(x_p u_n + y_p v_n)}; n = 1,...,N, p = 1,...,P\right\}$ and $\mathbf{F} = \{F(u_n, v_n); n = 1,...,N\}$, are the measurement matrix and the pattern samples vector, respectively, $k = 2\pi/\lambda$ is the wave number, λ being the wavelength, $u_n = \sin\theta_n \cos\phi_n$ and $v_n = \sin\theta_n \sin\phi_n$, $n = 1,...,N$, are the direction cosines, N being the number of pattern samples. In the following, according to the above formulation, two different inverse problems related to array processing, namely the diagnosis of array faulty elements and the design of unconventional clustered array architectures, are described and formulated in order to allow the use of *CS* in solve the arising inverse problems.

2.1. CS-based Diagnostic of Planar Phased Arrays

Let us assume that N_f elements of the array under test (*AUT*) are failed and the goal is to precisely estimate such failures starting from the knowledge of N measured samples of the respective far-field pattern. The vector of the *AUT* excitations can be mathematically described as

$$\mathbf{w}^{AUT} = diag(\mathbf{D})\mathbf{w} \tag{2}$$

where $\mathbf{D}$ is an $P \times P$ diagonal matrix whose diagonal values are equal to $d_{pp} = \alpha_p$, $p = 1,...,P$, where if $\alpha_p = 1$ the p-th element is correctly working, while if $\alpha_p < 1$ the p-th element is failed ($\alpha_p = 0$ if the p-th element is totally failed). The radiated *AUT* far-field patter is then acquired in N sampling points $\tilde{\mathbf{F}} = \{\tilde{F}(u_n, v_n); n = 1,...,N\}$ as:

$$\tilde{\mathbf{F}} = \mathbf{A}\mathbf{w}^{AUT} + \mathbf{e} \tag{3}$$

where $\mathbf{e} = \{e(u_n, v_n); n = 1,...,N\}$ is the vector of the measurement zero-mean Gaussian noise. Assuming the knowledge of the reference, fault-free array pattern $\mathbf{F}$, the difference between the *AUT* array pattern and the one radiated by the reference array, $\Delta\mathbf{F} = \mathbf{F} - \tilde{\mathbf{F}}$, can be obtained as the pattern radiated by an array fed by the differential weights $\Delta\mathbf{w} = \mathbf{w} - \mathbf{w}^{AUT}$. According to the above formulation, the addressed failure diagnosis problem is then reduced to the estimation of $\Delta\mathbf{w}$ starting from a set of N field measurements $\tilde{\mathbf{F}}$ satisfying the following relation

$$\Delta\mathbf{F} = \mathbf{A}\Delta\mathbf{w} + \mathbf{e} \tag{4}$$

It is worth noting here that if $N_f << P$, thanks to the above formulation, the vector $\Delta\mathbf{w}$ turns out to be sparse. Consequently, thanks to the linear relationship between the measured far-field samples, $\tilde{\mathbf{F}}$

and the differential weights vector, $\Delta\mathbf{w}$, the inverse problem (4) can be solved using the single-task Bayesian *CS* (*ST-BCS*) propsed in [22] for linear layouts, after properly transforming the complex matrix equation into a real-one [2][3].

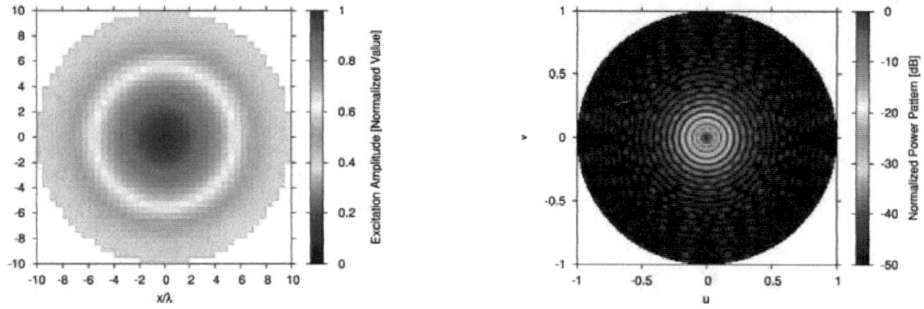

(a) (b)

Figure 1. *Numerical Examples (ST-BCS,* $P = 1264$, $N = 1257$ $N_f = 152$*)* – (a) Array excitation amplitude and (b) the array normalized power pattern.

2.2. Design of Clustered Linear Arrays by means of a TV-CS Strategy

Dealing with the design of clustered linear arrays by means of a CS-based scheme, let us formulate the synthesis problem as a pattern matching with respect to an available reference fully-populated (i.e. non-clustered) array. Let us consider a linear array of *P* elements placed along the x-axis, contiguously partitioned into *Q* sub-arrays. The vector of the equivalent array weights $\overline{\mathbf{w}} = \{\overline{w}_p; p = 1,...,P\}$ contains the P equivalent excitations coefficients expressed as:

$$\overline{w}_p = \sum_{q=1}^{Q} w_q^{SUB} \delta_{c_p q} \quad p = 1,...,P \tag{5}$$

where in eq. (5) w_q^{SUB} $q = 1,...,Q$ are the complex excitation weights of the Q sub-arrays, $\mathbf{c} = \{c_p \in [1, Q], p = 1,..., P\}$, is the clustering vector that univocally defines the sub-array membership of the *p-th* element to the *q-th* sub-arrays by means of Q integer numbers, and $\delta_{c_p q}$ is the Kronecker delta function ($\delta_{c_p q} = 1$ if $c_p = q$, otherwise $\delta_{c_p q} = 0$). Let us assume that the reference array far-field pattern $\mathbf{F}$ is known, the goal is to estimate the equivalent array excitations $\overline{\mathbf{w}}$ satisfying the pattern matching condition

$$\sum_{n=1}^{N} \left| F(u_n) - \sum_{p=1}^{P} \overline{w}_p e^{jkx_p u_n} \right|^2 \leq \varepsilon \tag{6}$$

with ε reasonably small. It is worth noting here that if the array is partitioned into $Q << P$ contiguous clusters, the vector $\overline{\mathbf{w}}$ turns out to be a constant function with few discontinuities. Consequently the gradient of the vector $\overline{\mathbf{w}}$ defined as:

$$\nabla\overline{\mathbf{w}} = \{\nabla_p \overline{\mathbf{w}} = w_p - w_{p-1}\} \tag{7}$$

turns out to be a sparse vectors with all zero entries, except for the p indexes corresponding to the elements that are located at the border of the clusters. Accordingly, the pattern matching problem (6) can be solved using a *CS*-based approach. More in detail the Total Variation *CS* (*TV-CS*) method [6] can be directly applied in order to retrieve TV-sparse solutions effectively fitting the proposed formulation. With respect to the state-of-art *CS*-based technique [26] the proposed method does not need to define a-priori the size of the clusters, allowing a complete exploration of the existing solutions.

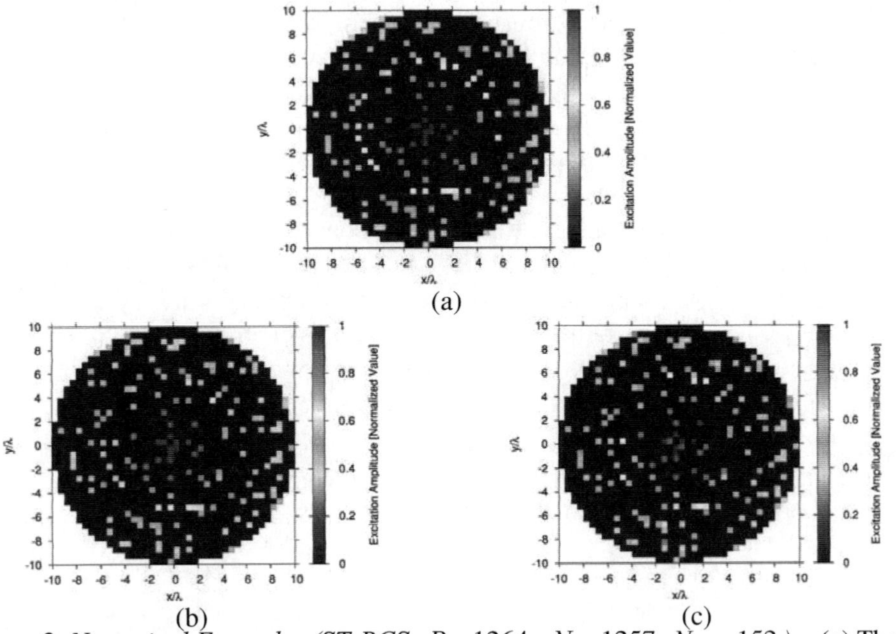

Figure 2. *Numerical Examples (ST-BCS,* $P = 1264$, $N = 1257$ $N_f = 152$*)* – (a) The actual failures and (b)(c) the *CS* estimated failures considering noisy field samples with (b) $SNR = 40$ [dB] and (c) $SNR = 20$ [dB].

3. Numerical Examples

As a first numerical example of array diagnosis using a *CS* approach, a planar circular array of $P = 1264$ isotropic sources equally spaced by $d_x = d_y = 0.5\lambda$ is considered. The excitations amplitude have been obtained according to a Taylor distribution ($\overline{n} = 3$, $SLL = -25$ [dB]) as shown in Fig. 1(a), radiating the far-field power pattern reported in Fig. 1(b). The test case considers an *AUT* in which a subset of elements is totally failed while all the other elements are correctly working (i.e. $\alpha_p = 0$ or $\alpha_p = 1$, $p = 1,...,P$). The number of failed elements is equal to $N_f = 152$ (i.e. a percentage of failure equal to $\chi = 12\%$) as shown in Fig. 2(a), reporting the differential weights $\Delta\mathbf{w}$. Starting from the knowledge of $N = 1257$ noisy samples of the far field radiated by the *AUT*, the failure configuration has been estimated using the *ST-BCS* solver [22], and reported in Fig. 2(b)-(c). As can be seen, the *CS* method allows to estimate all the $N_f = 152$ failures with small deviations in terms of the amplitude for both the considered noisy levels [i.e. *SNR*=20 [dB], Fig. 2(b) and SNR=40 [dB], Fig. 2(c)]. Of course the percentage of failures that can be reliably estimated is limited, due to the sparseness of the solution enforced in solving eq. (4). A maximum number of failures recovered by the proposed ST-BCS based method, has been estimated in the interval $\chi_{\max} = [16\%, 20\%]$ if reliable reconstruction must be guaranteed. In order to improve such limitation, the use of the *MT-BCS* approach [10] will be considered as a possible extension of the methodology, exploiting as instance, the correlation of the real and imaginary parts of the measured pattern values. It is worth noting that the proposed approach, according to the formulation reported in Sec. 2, can handle only pixel-like failures, consequently, future extensions will consider non-pixel like failures, such as clusters of failed elements complying with unconventional tiled architectures [27].

As a second numerical example a linear array of $P = 20$ ideal isotropic elements, laying in the x-axis equally spaced by $d_x = 0.5\lambda$ is considered. The goal is to find a clustered solution that minimize the matching with the reference array reported in Fig. 3(c) (Taylor, $\bar{n} = 6$, $SLL = -20$ [dB]). Figure 3(a) shows the clustering configuration corresponding to the solution provided by the *TV-CS* solver [Fig. 3(b)], which allows to reduce the number of control points to $Q = 13$. Figure 3(c) compares the power pattern of the reference array with the *TV-CS* synthesized pattern. As can be seen the *TV-CS* pattern is very close to the reference with a non negligible simplification of the feeding network (i.e. a reduction of $\chi = 35\%$ of amplifiers). Higher reduction percentages can be achieved when considering larger arrays. As instance, Fig. 4 shows the case of a $P = 100$ elements array. Two different solutions, selected according to two different sampling rates, are shown in Fig. 4(a) ($N = 90$) and Fig. 4(b) ($N = 200$), characterized by $Q = 25$ and $Q = 33$ clusters, respectively. The clustered amplitude coefficients are compared in Fig. 4(c) while the respective power patterns are reported in Fig. 4(d). As can be notice, the power pattern radiated by the $Q = 33$ sub-arrays solution shows negligible deviations from the optimal pattern, with a final reduction of $\chi = 65\%$ of control points. Instead, the $Q = 25$ solution allows a further simplification of the feeding network ($\chi = 75\%$) accepting the presence of slightly higher lobes in the side-lobe region [Fig. 4(d)]. Future extensions of the proposed approach include the possibility to handle planar array structures and, by exploiting suitable basis functions, to optimize the partitioning of tile-based modular arrays [27].

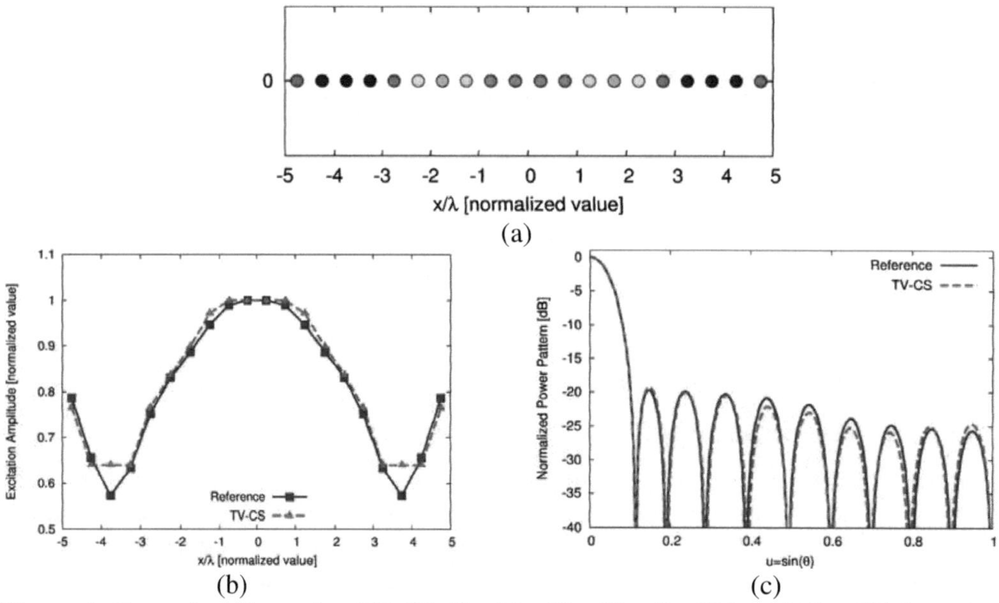

Figure 3. *Numerical Examples (TV-CS,* $P = 20$, $N = 90$, $Q = 13$*)* – (a) The *CS* clustering configuration, (b) the reference and *TV-CS* excitations amplitudes and (c) the normalized reference and *TV-CS* synthesized power patterns.

4. Conclusions

In this work, a review of the main *CS*-based methodologies applied to inverse problems arising in the design and processing of antenna arrays, is reported. Two instances of inverse problems have been formulated in order to comply with the *CS* theoretical requirements, namely the diagnosis of failures in planar arrays and the design of contiguously clustered arrays, discussing the limitations and the future trends of both techniques. Finally two illustrative examples are reported, validating the effectiveness of the proposed method.

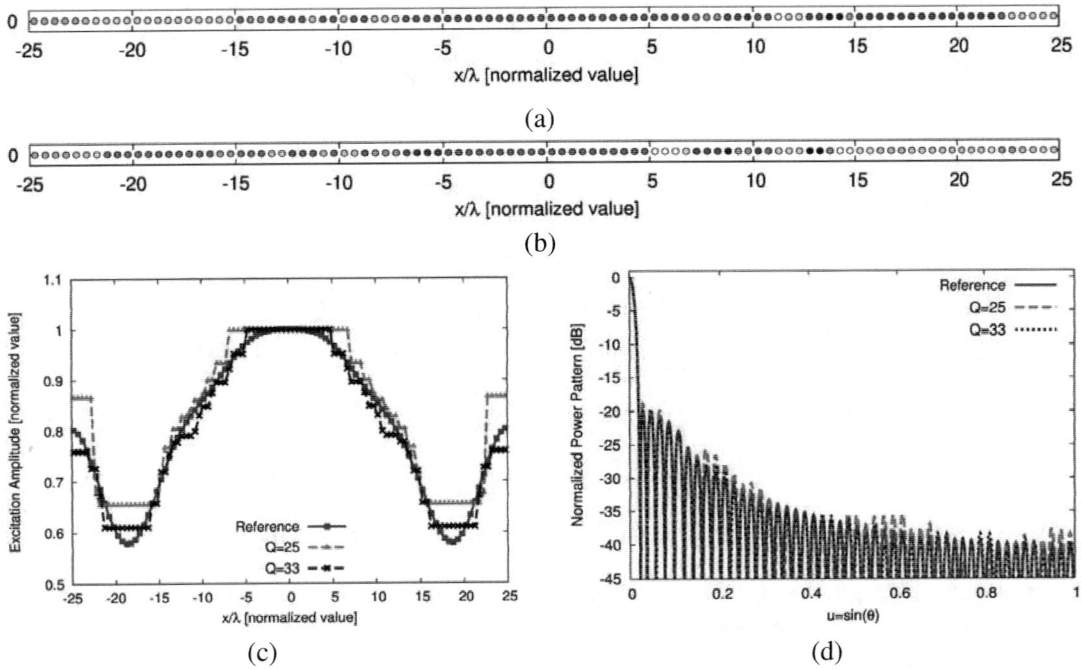

Figure 4. *Numerical Examples (TV-CS, $P = 100$)* – (a)(b) The *CS* clustering configurations of two TV-CS solutions with (a) $Q = 25$ and (b) $Q = 33$ clusters. (c) the reference and *TV-CS* excitations amplitudes and (d) the normalized reference and *TV-CS* synthesized power patterns.

Acknowledgments
This work benefited from the networking activities carried out within the Càtedra de Excelencia UC3M-Santander funded by the Universidad Carlos III de Madrid (Spain) (Prof. A. Massa).

References
[1] Candes E J and Wakin M B 2008 An introduction to compressive sampling *IEEE Signal Process. Mag.* **25** 21-30
[2] Massa A, Rocca P and Oliveri G 2015 Compressive sensing in electromagnetics - A review *IEEE Antennas Propag. Mag.* **57** 224-38
[3] Oliveri G, Salucci M, Anselmi N and Massa A 2017 Compressive sensing as applied to inverse problems for imaging: theory, applications, current trends, and open challenges *IEEE Antennas Propag. Mag.* **59** 34-46
[4] Candes E J 2008 The restricted isometry property and its implications for compressed sensing *C. R. Math.* **346** 589-92
[5] Pan L, Chen X, and Yeo S P 2012 A compressive-sensing-based phaseless imaging method for point-like dielectric objects *IEEE Trans. Antennas Propag.* **60** 5472–5
[6] Oliveri G, Anselmi N and Massa A 2014 Compressive sensing imaging of non-sparse 2D scatterers by a total-variation approach within the Born approximation *IEEE Trans. Antennas Propag.* **62** 5157-70
[7] Anselmi N, Oliveri G, Salucci M and Massa A 2015 Wavelet-based compressive imaging of sparse targets *IEEE Trans. Antennas Propag.* **63** 4889-900
[8] Anselmi N, Oliveri G, Hannan M A, Salucci M and Massa A 2017 A Color compressive

sensing imaging of arbitrary-shaped scatterers *IEEE Trans. Microw. Theory Techn.* **65** 1986-99

[9] Zhang W, Li L and Li F 2011 Reducing the number of elements in linear and planar antenna arrays with sparseness constrained optimization *IEEE Trans. Antennas Propag.* **59** 3106–11

[10] Oliveri G, Carlin M and Massa A 2012 Complex-weight sparse linear array synthesis by Bayesian Compressive Sampling *IEEE Trans. Antennas Propag.* **60** 2309-26

[11] Viani F, Oliveri G and Massa A 2013 Compressive sensing pattern matching techniques for synthesizing planar sparse arrays *IEEE Trans. Antennas Propag.* **61** 4577-87

[12] Carlin M, Oliveri G and Massa A 2015 Hybrid BCS-deterministic approach for sparse concentric ring isophoric arrays *IEEE Trans. Antennas Propag.* **63** 378-83

[13] Bencivenni C, Ivashina M V, Maaskant R and Wettergren J 2016 Synthesis of maximally sparse arrays using compressive sensing and full-wave analysis for global earth coverage applications *IEEE Trans. Antennas Propag.* **64** 4872–7

[14] D'Urso M, Prisco G and Tumolo R M 2016 Maximally sparse, steerable, and nonsuperdirective array antennas via convex optimizations *IEEE Trans. Antennas Propag.* **64** 3840-9

[15] Fuchs B and Rondineau S 2016 Array pattern synthesis with excitation control via norm minimization *IEEE Trans. Antennas Propag.* **64** 4228-34

[16] Oliveri G, Bekele E T, Robol F and Massa A 2014 Sparsening conformal arrays through a versatile BCS-based method *IEEE Trans. Antennas Propag.* **62** 1681-9

[17] Pinchera D, Migliore M D, Schettino F, Lucido M and Panariello G 2018 An effective compressed-sensing inspired deterministic algorithm for sparse array synthesis *IEEE Trans. Antennas Propag.* **66** 149-159

[18] Model D and Zibulevsky M 2006 Signal reconstruction in sensor arrays using sparse representations *Signal Process.* **86** 624–38

[19] Rocca P, Hannan M A, Salucci M and Massa A 2017 Single-snapshot DoA estimation in array antennas with mutual coupling through a multi-scaling BCS strategy *IEEE Trans. Antennas and Propag.* **65** 3203-13

[20] Carlin M, Rocca P, Oliveri G, Viani F and Massa A 2013 Directions-of-arrival estimation through Bayesian Compressive Sensing strategies *IEEE Trans. Antennas Propag.* **61** 3828-38

[21] Carlin M, Rocca P, Oliveri G and Massa A 2013 Bayesian compressive sensing as applied to directions-of-arrival estimation in planar arrays *J. Elect. Comput. Eng.* **245867**

[22] Oliveri G, Rocca P and Massa A 2012 Reliable diagnosis of large linear arrays - A Bayesian Compressive Sensing approach *IEEE Trans. Antennas Propag.* **60** 4627-36

[23] Migliore M D 2013 Array diagnosis from far-field data using the theory of random partial Fourier matrices *IEEE Antennas Wireless Propag. Lett.* **12** 745–8

[24] Ince T and Ogucu G 2016 Array failure diagnosis using nonconvex compressed sensing *IEEE Antennas Wireless Propag. Lett.* **15** 992-5

[25] Fuchs B, Le Coq L and Migliore M D 2016 Fast antenna array diagnosis from a small number of far-field measurements *IEEE Trans. Antennas Propag.* **64** 2227-35

[26] Oliveri G, Salucci M and Massa A 2016 Synthesis of modular contiguously clustered linear arrays through a sparseness-regularized solver *IEEE Trans. Antennas Propag.* **64** 4277-87

[27] Anselmi N, Rocca P, Salucci M and Massa A 2017 Irregular phased array tiling by means of analytic schemata-driven optimization *IEEE Trans. Antennas Propag.* **65** 4495-510

IOP Conf. Series: Journal of Physics: Conf. Series **1131** (2018) 012005 doi:10.1088/1742-6596/1131/1/012005

Computational methods for wireless structural health monitoring of cultural heritages

M Bertolli[1], M Donelli[1], A Massa[1,2], G Oliveri[1], A Polo[1], F Robol[1], L Poli[1], A Gelmini[1], G Gottardi[1], M A Hannan[1], L T P Bui[1], P Rocca[1], C Sacchi[1], F Viani[1], T Moriyama[3], T Takenaka[3] and M Salucci[1]

[1] ELEDIA Research Center (ELEDIA@UniTN - University of Trento)
Via Sommarive 9, I-38123 Trento, Italy
[2] ELEDIA Research Center (ELEDIA@UC3M - Universidad Carlos III de Madrid)
Avenida de la Universidad 30, 28911 Leganés, Madrid – Spain
[3] ELEDIA Research Center (ELEDIA@UniNAGA - University of Nagasaki)
852-8521, Nagasaki, Japan

E-mail: andrea.massa@unitn.it

Abstract. The structural health monitoring of cultural heritages is addressed in this paper. The arising inverse problem is solved through the Learning-by-Examples (*LBE*) paradigm, exploiting data collected by a Wireless Sensor Network (*WSN*). More in detail, low-cost and low-size sensing devices are spread over the scenario to be monitored, allowing environmental data as well as acceleration and vibration information to be acquired and processed by means of a Support Vector Machine (*SVM*) in order to detect the presence/absence of a damage in the monitored structure. The proposed approach has been preliminary validated in a laboratory-controlled environment, demonstrating promising performance.

1. Introduction

Structural Health Monitoring (*SHM*) is a process aimed at the detection and characterization of structural damages in structures of any kind. In particular, literature reports several examples of local Non-Destructive (*ND*) and Destructive Testing (*DT*) methodologies for structural health detection as applied from civil to aerospace engineering structures. In the last years, in fact, there has been an increasing interest in such a science, especially considering the high potential life-safety and the economic benefits, which can be obtained thanks to the understanding of the actual performance and future life of a structure. In addition, *SHM* has also been boosted thanks to the technological evolution in pervasive computing and computer science, which has determined the ability of manipulating big amount of data acquired by several distributed sensors, by applying complex methodological systems to achieve advanced *ND* structural health estimations [1]-[7].

The *SHM* process generally involves the observation of a system over time, which may be performed periodically or continuously in order to sample the dynamic evolution of the environmental characteristics, which may somewhere damage or deteriorate the structure itself. Such systems are indeed oriented to the detection and forecasting of structural damages, potentially enabling the scheduling of preventive maintenance operations. On the other hand, *SHM* is also used to detect structural damages in case of extreme events such as earthquakes [8]. As an example, Japan is one of the most earth-quakes zone in the world. The costs for structure maintenance is very high, hence a high interest in *SHM* applications has arisen in the last years.

To this end, Wireless Sensors Networks (*WSN*s) represent a widely used technology, thanks to the provided features such as low-cost and easy installation, which allowed a rapid increase in the use of wireless-based systems as an alternative to wired-based networks [5].

In order to perform the structural health analysis and solve the related inverse problem (i.e., by inferring the status of the monitored structure starting from *WSN* pervasive measurements), some tests have to be performed, thus acquiring useful information to be processed for the damage investigation. In such a framework, destructive testing represents a set of investigative techniques, which determine the material mechanical characteristic, such as strength, hardness or toughness, through material breaking. The advantage of this kind of tests, is related to the possibility to determine material features, which cannot be identified with non-destructive tests. At the same time, this advantage represents also one of the main drawbacks of such techniques since the structure to be monitored has to be totally or partially impaired. Accordingly, destructive techniques become more economic for mass produced objects.

SHM is also referred to a wide variety of non-destructive testing [9]. To establish the health status of a structure and to determine its reliability and safety, continuous inspections should be performed. Traditionally, visual inspection or simpler measurements are performed by a technician, but this kind of analyses may not be sufficient since only information about the superficial health status can be collected, neglecting potential internal damages. Hence, sometimes non-destructive testing has to be paired with destructive testing applied to samples of the monitored structures, which have to be removed from the original structure. In this way, destructive testing is conducted in laboratory on the structure samples, in order to evaluate the mechanical properties. As it may be clear, these operations may not always be feasible, especially considering the cultural heritage works.

Nevertheless, the evolution of electronics and computer science, and an increasing interest in these applications, have allowed the study and definition of general approaches in non-destructive techniques, such as systems based on the study of the propagation of mechanical waves [10] or of an induced electromagnetic field inside the monitored structure, or systems based on radar and radio frequency analyses [11][12].

SHM processes can be articulated into four main levels, namely damage detection, damage localization, damage quantification, prognosis on the remaining life time.

Damage detection is the process in charge of identifying the damage in the monitored structure, in general comparing current conditions with respect to a reference status, such as a model or experimental measure [13]. In order to make the discrimination precise and reliable, it is necessary to suppress environmental influence, filtering data from undesired measurement noise and considering additional data acquisition to compensate external influence.

Damage localization, instead, represents the statistical process used to estimate the actual position of the structural damage on the basis of the acquired data. Damage localization is often based on the wave propagation theory, as applied to materials, also exploiting probabilistic approaches to estimate the position of structural damages [14]. In the literature, two main techniques are considered, respectively based on mechanical and electromagnetic waves. Single or multiple damages could be recognized by the use of particular techniques for data manipulation [15]. In addition, environmental condition, especially temperature, impact on the physical characteristics of materials, which actually represent the media where mechanical and electromagnetic waves are propagating. For this reason, also damage localization algorithms need proper data compensation to filter undesired external influences.

Damage quantification approaches allow to estimate the structural degradation and damages as a combination of multiple measures and evaluations. Such a process can be recast as a decision-making procedure, which has to outcome quantitative and detailed damage information [16]. A crucial aspect is therefore represented by the *goodness* of the acquired data, which may improve the actual result obtained from the damage quantification process. As a comparison, the damage detection process requires lower complexity to estimate a pattern change with respect to a reference situation. On the contrary, damage quantification complexity further grows in order to achieve accurate quantitative damage estimation, considering both required sensors and processing methodology. As an example, damage quantification can be computed by using high performance decomposition methods, which allow a high resolution in damage estimation [17].

After damage detection, localization, and quantification the remaining life time of the monitored structure can be estimated [18]. In general, depending on the structure, a number of damages due to climate solicitation and mechanical stress could be supported [19]. For this reason, an *SHM* system has to properly prevent failures in decision making, thus giving correct suggestions on decide to repair, upgrade or dismantle it. In [20], the authors introduce a two-steps approach. The first step is focused on the detection of low entity damages (called micro-cracks). Multiple micro-cracks could be accumulated by materials. Thus, a probabilistic approach is used in the second step, in order to predict the onset of high entity damages (called macro-cracks), which could permanently damage the materials composing the structure, with potential severe loss of stability and safety.

In general, large set of sensors may be deployed on real sites thus improving the pervasiveness of the monitoring system. In such a scenario, sensors could be of heterogeneous nature, in order to achieve the objective of the structure health estimation. In detail, some sensors deployed at the site should be devoted to measure the actual structure properties, while other sensors may be employed in order to acquire useful information to compensate material properties changes due to the environment conditions.

Cultural heritage monitoring application is mainly focused on artwork maintenance, by exploiting distributed sensing for preserving them. Therefore, in order to preserve paintings, historical structures, frescos and artworks, pervasive monitoring systems can be exploited in order to continuously monitor the cultural heritage and prevent potential damages [21]. In [22], a monitoring system for wooden artworks is proposed with environmental and mechanical measurements. A predictive simulator of stress is also proposed to estimate wood deformation due to environmental conditions.

Therefore, considering the *SHM* for cultural heritage monitoring, this work is aimed at proposing a wireless monitoring system for damage detection, by exploiting the heterogeneous data acquired by distributed *WSN* sensing devices and processed by a machine learning methodology [2]-[7], [23]-[25] in order to determine if potential damages are arising in the monitored structure.

2. System Architecture

The proposed system for *SHM* of cultural heritages is based on the *WSN* paradigm, thus allowing the system to be spread even over large environments, such as museums or exhibition areas. Therefore, a set of N sensing devices can be deployed around the scenario in order to sense the heterogeneous parameters of interest at different positions. A data collection device, commonly called gateway node (*GW*) or data concentrator, is used as an interface between the wireless sensor network and the data analysis system.

The data measured by the sensor network, once received by the control unit, are properly acquired and the reception timestamp is appended to the data stream for synchronization purposes, before storing such information in a local or remote database. The choice of storing data into a local or remote database is mainly motivated by the ease of management of big data streams and in turn the ease of recalling whatever subset of saved data for subsequent processing. A multiple database approach may be considered, as well, in order to guarantee data on remote machines and allow the processing of information though computer clouding approaches.

The wireless sensing node is based on the TinyNode device, properly connected to a sensor interfacing board developed ad-hoc for the communication with the selected sensing technologies modules. In particular, the wireless sensing node can be equipped with a temperature and humidity sensor (SHT11 model), a piezoelectric transducer (BM15015-06HC model), and a high resolution three-axis digital accelerometer (LIS3LV02 model).

IOP Conf. Series: Journal of Physics: Conf. Series **1131** (2018) 012005 doi:10.1088/1742-6596/1131/1/012005

Figure 1 shows a wireless sensor device as mounted in the selected packaging, which contains the TinyNode device, the sensor interfacing board and the batteries for powering the device. The piezoelectric transducer as well as the accelerometer can be installed outside the packaging for direct contact with the monitored structure.

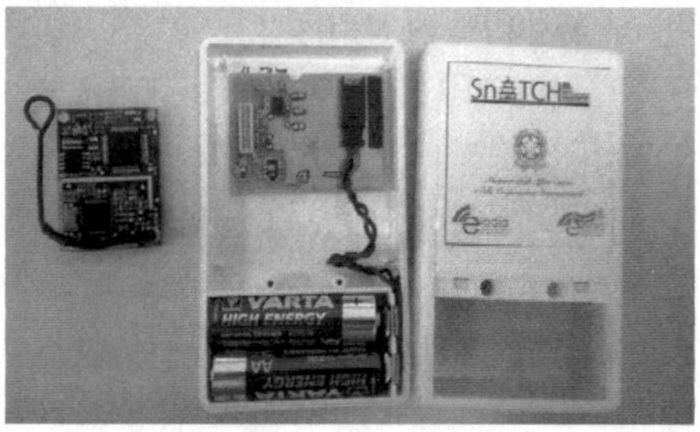

Figure 1. Detail of the wireless sensor device interface board for sensor equipment.

The sensing nodes communicate among the wireless network in the *ISM* 868 MHz band. This design choice is driven by the low scattering given by the actual operating frequency, which allows the ease deployment of the monitoring system in large environments, without the need of forwarding nodes and/or procedures to allow each sensing device to communicate with the *GW* node.

The acquired data are representative of the health status of the monitored structure. In particular, a variation of the sensed parameters is expected in case of a damage, if compared with the damage absence scenario data. Therefore, as a preliminary approach, the damage detection problem is recast as a binary classification problem solved by means of a support vector machine (*SVM*) algorithm [24].

During a preliminary off-line phase, a database of input/output (*I/O*) pairs (i.e., presence/absence of damages in the structure vs. sensed parameters) is setup in order to allow the *SVM* algorithm in determining the decision function, successively used during the on-line phase to perform real-time predictions of the structure health (i.e., detecting the presence or absence of a damage).

More in detail, a set of *K I/O* pairs of data $\left\{\boldsymbol{\xi}^{(k)}, \chi\left(\boldsymbol{\xi}^{(k)}\right); k = 1,\ldots,K\right\}$ is generated during the off-line *SVM* training procedure to model the relationship between sensed data $\boldsymbol{\xi}^{(k)} = \left\{\left(T_n^{(k)}, H_n^{(k)}, \underline{A}_{n,x}^{(k)}, \underline{A}_{n,y}^{(k)}, \underline{A}_{n,z}^{(k)}, \underline{V}_n^{(k)}\right); n = 1,\ldots,N\right\}$, where $T_n^{(k)}$ and $H_n^{(k)}$ are respectively the temperature and relative humidity for the n-th node in the *WSN*, $\underline{A}_{n,q}^{(k)} = \left\{a_{n,q,i}^{(k)}; i = 1,\ldots,I\right\}$, $q = \{x, y, z\}$, $a_{n,q,i}^{(k)}$ being the q-th acceleration component measured by the n-th node at the i-th time instant $t_i \in [0, \tau], i = 1,\ldots,I$, τ being the acquisition time window, $\underline{V}_n^{(k)} = \left\{v_{n,i}^{(k)}; i = 1,\ldots,I\right\}$ where $v_{n,i}^{(k)}$ is the vibration measured by the n-th node at the i-th instant, and the health status of the structure given by $\chi\left(\xi^{(k)}\right) = +1$ when a damage is present, $\chi\left(\xi^{(k)}\right) = -1$ in case of absence of damages.

Therefore, the information carried out by such training set is exploited by the *SVM* algorithm in order to define the following decision function

$$\hat{\chi}(\boldsymbol{\xi}) = \mathrm{sgn}\left(\sum_{k=1}^{K} \chi\left(\boldsymbol{\xi}^{(k)}\right) \alpha^{(k)} \, \Omega\left(\boldsymbol{\xi}^{(k)}, \boldsymbol{\xi}\right) + b\right) \tag{1}$$

used during the on-line working of the algorithm for predicting the structure health status.
In Eq. (1), $\mathrm{sgn}(.)$ is the sign function, $0 \le \alpha \le C$ are the Lagrange multipliers with C being the so-called penalty factor, b is a bias term [23]-[25], while $\Omega\left(\boldsymbol{\xi}^{(k)}, \boldsymbol{\xi}\right) = \exp\left(-\gamma \left\|\boldsymbol{\xi}^{(k)} - \boldsymbol{\xi}\right\|^2\right)$ is the radial basis function (*RBF*) kernel function.

IOP Conf. Series: Journal of Physics: Conf. Series **1131** (2018) 012005 doi:10.1088/1742-6596/1131/1/012005

3. Experimental Validation

The proposed system has been arranged in a laboratory-controlled scenario at the ELEDIA@UniTN laboratories, University of Trento, Italy. In particular, a set of $N = 10$ wireless sensing devices have been spread around the laboratory areas, as shown in Figure 2, in order to allow the validation of both the wireless monitoring system capabilities, testing the communication performance in different conditions representative of potential real applications (e.g., museum exhibition, ancient buildings characterized by thick walls that impact on the communication range capability), as well as the performance of the *LBE*-based strategy for damage detection.

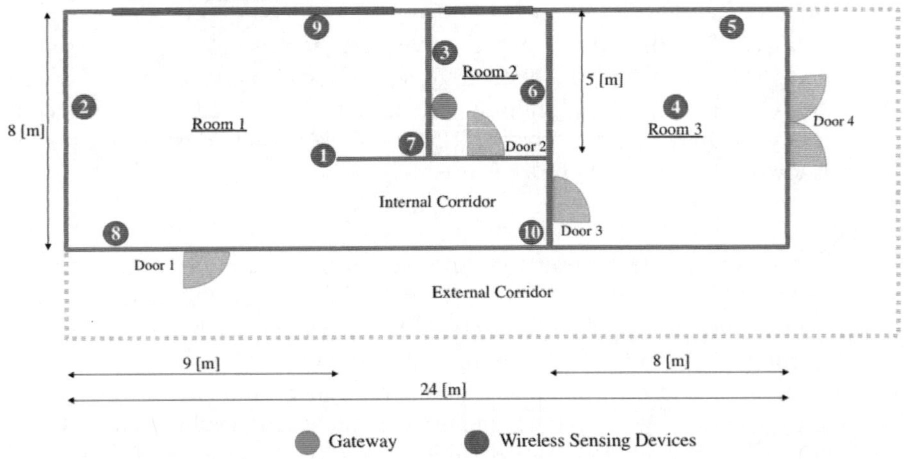

Figure 2. Experimental validation scenario, wireless sensing devices position.

Each device has been equipped with the sensing modules described in the previous section and in particular an acquisition time $\tau = 1[\text{sec}]$ for the acceleration and vibration modules has been considered. Experimental testing is currently under development in order to define an exhaustive set of *I/O* pairs of data to be used for the off-line training of the *SVM* in order to obtain good generalization capabilities of the proposed detection system. In particular, different structures are under analysis to characterize their properties and behavior under different environmental and stress conditions.

Figure 3. Experimental validation, wireless sensing device applied to wood structure.

Some preliminary results have demonstrated the potentialities of the proposed system in structural damage detection on the basis of non-destructive analysis performed with low cost and reduced size wireless sensing devices.

More in detail, a preliminary training set consisting of $K = 100$ *I/O* samples has been generated by dropping objects of different weights onto some monitored wood structures (e.g., representative of a painting frame, Figure 3) in order to simulate anomalous vibrations, deformations, and mechanical stresses of different nature. Moreover, a test/validation set of $M = 50$ samples, evenly distributed (as for training data) between "damaged" and "undamaged" labels, has been considered to assess the prediction accuracy of the trained *SVM*. Concerning the optimal setting of the parameters of the adopted binary classifier, a standard 5-fold cross-validation approach has been employed on the available training data [2][25], the optimal values being determined as $C = 10^2$ and $\gamma = 10^{-1}$ for the penalty factor and the *RBF* kernel width, respectively. Under such a configuration, the time required to train the *SVM* (once training data have been experimentally collected) has been lower than 10 seconds using a standard laptop and a non-optimized software implementation.

The proposed system showed an appreciable capability in detecting the presence or absence of the applied stress to the wood structure, allowing an acceptable degree of discrimination (i.e., a classification accuracy of about $\Gamma = 67\%$) between class -1 ("absence of a damage") and class $+1$ ("presence of a damage"). Of course, higher detection accuracies are envisaged in the next experimental campaigns by considering a better and more rigorous approach to generate training data, as well as by increasing the number of training samples, *K*. Moreover, it should be remarked that, differently from many state-of-the-art works [8][9][13], the developed monitoring system is based on the exploitation of low-cost sensing devices.

4. Conclusions

SHM represents an interesting technique for cultural heritage monitoring with specific focus on maintenance and preservation of buildings and artworks. In particular, the damage detection has been formulated as an inverse problem, solved through a *LBE*-based approach applied to the data acquired by a distributed monitoring system able to measure different heterogeneous physical parameters. The experimental validation is currently under development, and preliminary results demonstrated the potentialities of the proposed system. It is understood that such results have been obtained with inexpensive off-the-shelf hardware and considering sub-optimal approaches for building the training set, testing the on-line accuracy of the system, and calibrating it in order to enhance robustness to noise and environmental changes that are expected to occur in real-world deployments. Current and future work will be devoted to additional analysis and testing of several structures in order to improve the generalization capabilities of the system. Moreover, an in-depth analysis of the sensitivity of the *SVM* accuracy to relevant parameters such as C, γ, and K will be object of careful investigation. Finally, the use of advanced dimensionality reduction techniques (i.e., feature-selection/feature-extraction) will be investigated in order to determine the most relevant input features/sensed parameters, improve the overall detection and monitoring accuracy, as well as reduce the number of necessary training samples, mitigating the negative effects of the so-called "*curse of dimensionality*" [2].

Acknowledgment

This work has been partially supported by the Italian Ministry of Foreign Affairs and International Cooperation, Directorate General for Cultural and Economic Promotion and Innovation within the SNATCH Project (2017-2019).

References

[1] Chang P C and Liu S C 2003 Recent Research in Nondestructive Evaluation of Civil Infrastructures *J. Mater. Civil Eng.* **14** 298-304

[2] Massa A, Oliveri G, Salucci M, Anselmi N and Rocca P 2018 Learning-by-examples techniques as applied to electromagnetics *J. Electromagn. Waves Appl.* **32** 516-541

[3] Donelli M, Viani F, Rocca P and Massa A 2009 An innovative multi-resolution approach for DoA estimation based on a support vector classification *IEEE Trans. Antennas Propag.* **57** 2279-2292

[4] Viani F, Migliore M D, Polo A, Salucci M and Massa A 2018 Iterative classification strategy for multi-resolution wireless sensing of passive targets *Electron. Lett.* **54** 101-103

[5] Viani F, Robol F, Polo A, Rocca P, Oliveri G and Massa A 2013 Wireless architectures for heterogeneous sensing in smart home applications - concepts and real implementations *Proc. IEEE* **101** 2381-2396

[6] Salucci M, Vrba J, Merunka I and Massa A 2017 Real-time brain stroke detection through a learning-by-examples technique - an experimental assessment *Microw. Opt. Technol. Lett.* **59** 2796-2799

[7] Salucci M, Anselmi N, Oliveri G, Calmon P, Miorelli R, Reboud C and Massa A 2016 Real-time NDT-NDE through an innovative adaptive partial least squares SVR inversion approach *IEEE Trans. Geosci. Remote Sens.* **54** 6818-6832

[8] Ferrar C R and Worden K 2010 An introduction to Structural Health Monitoring *New Trends in Vibration Based Structural Health Monitoring, CISM Courses and Lectures* **520**

[9] Chang P C, Flatau A and Liu S C 2003 Review Paper: Health Monitoring of Civil Infrastructure *Struct. Health. Monit.* **2** 257-267

[10] Giancoli D C 2009 Physics for scientist & engineers with modern physics, 4th ed *Upper Saddle River, N.J. Pearson Prentice Hall*

[11] Perez-Gracia V, Di Capua D, Gonzalez-Drigo R, Caselles O, Pujades L G and Salinas V 2010 GPR resolution in Cultural Heritage applications *Proceedings of the XIII Internarional Conference on Ground Penetrating Radar* 1-5

[12] Salucci M, Oliveri G and Massa A 2015 GPR prospecting through an inverse-scattering frequency-hopping multifocusing approach *IEEE Trans. Geosci. Remote Sens.* **53** 6573-6592

[13] Fritzen C 2014 Vibration-Based Methods for SHM *North Atlantic Treaty Organization Science and Technology Organization*

[14] Neerukatti R K, Hensberry K, Kovvali N and Chattopadhyay A 2015 A novel probabilistic approach for damage localization and prognosis including temperature compensation *J. Intell. Material Syst. Struct.* **27** 592-607

[15] Undewood S S, Mayer J J and Adams D E 2015 Damage Localization in Composite Structures Using Nonlinear Vibration Response Properties *J. Vib. Acoust.* **137**

[16] Mesnil O, Leckey C A and Ruzzene M 2014 Instantaneous and local wavenumber estimation for damage quantification in composites *Struct. Health. Monit.* **14** 193-204

[17] Das S, Kyriakides I, Chattopadhyay A and Papandreu-Suppappola A 2009 Monte Carlo Matching Pursuit Decomposition Method for Damage Quantification in Composite Structure *J. Intell. Material Syst. Struct.* **20** 647-657

[18] Gobbato M, Conte J P, Kosmatka J B and Ferrar C R 2012 A reliability-based framework for fatigue damage prognosis of composite aircraft structures *Probabilist. Eng. Mec.* **29** 176-188

[19] Kohn S J, Maalej M and Quek S T 2004 Damage Quantification of Flexurally Loaded RC Slab Using Frequency Response Data *Struct. Health. Monit.* **3** 293-311

[20] Kulkarni S S and Achenbach J D 2008 Structural Health Monitoring and Damage Prognosis in Fatigue *Struct. Health. Monit.* **7** 37-49

[21] D'Amato F, Gamba P and Goldoni E 2012 Monitoring heritage buildings and artworks with Wireless Sensor Networks *IEEE Workshop on Environmental Energy and Structural Monitoring Systems* 1-6

[22] Allegretti O, De Vicenzi M, Uzielli L and Dionisi-Vici P 2013 Long-term hygromechanical monitoring of Wooden Objects of Art (WOA): A tool for preventive conservation *J. Cult. Herit.* **14** 191-164

[23] Ahmed S, Salucci M, Miorelli R, Anselmi N, Oliveri G, Calmon P, Reboud C and Massa A 2017 Real time groove characterization combining partial least squares and SVR strategies: application to eddy current testing *J. Phys. Conf. Ser.* **904** 1-8

[24] Ahmed S, Miorelli R, Salucci M and Massa A 2017 Real-time flaw characterization through learning-by-examples techniques: a comparative study applied to ECT *Electromagnetic Nondestructive Evaluation (XX)* **42** 228-235

[25] Cortes C and Vapnik V 1995 Support - vector networks *Mach. Learn.* 273–297

IOP Conf. Series: Journal of Physics: Conf. Series **1131** (2018) 012006 doi:10.1088/1742-6596/1131/1/012006

Processing GPR data with 2D Bayesian compressive sensing inverse scattering approaches

P Rocca[1], **G Gottardi**[1], **M Bertolli**[1], **F Robol**[1], **T Moriyama**[2] **and T Takenaka**[2]

[1] ELEDIA Research Center (ELEDIA@UniTN - University of Trento)
Via Sommarive 9, I-38123 Trento, Italy
[2] ELEDIA Research Center (ELEDIA@UniNAGA - University of Nagasaki)
852-8521, Nagasaki, Japan

E-mail: paolo.rocca@unitn.it

Abstract. In this work, the solution of the subsurface microwave imaging problem is addressed through innovative inverse scattering (*IS*) approaches. Towards this end, wide-band ground penetrating radar (*GPR*) measurements are processed by means of Bayesian Compressive Sensing (*BCS*)-based strategies integrated with frequency-hopping (*FH*) or multi-frequency (*MF*) inversion schemes. Thanks to the additional information provided by the frequency diversity of *GPR* data and the formulation of suitable *sparseness*-regularized methodologies faithful reconstructions are yielded under several operative conditions, as clearly indicated by selected numerical experiments.

1. Introduction

The development of inverse scattering (*IS*) techniques for solving the subsurface microwave imaging problem is nowadays a very attractive field of research given the wide set of civil and military applications. Within this context, ground penetrating radar (*GPR*) is one of the most versatile non-destructive prospecting tool thanks to many favorable features including easy deployment and low cost, making it the perfect candidate for cultural heritage monitoring, civil engineering, forensic investigations, through-the-wall imaging, as well as de-mining [1]-[10]. However, one of the most critical limitations when using *GPR* as probing instrument is the need for high user experience for the correct interpretation of the measured *radargrams*. To overcome such a limitation, several *IS* approaches have been developed in the recent years for the automatic processing of *GPR* data and the generation of easy-to-interpret qualitative and/or quantitative reconstructions of the probed domain [2]-[10]. Among the many proposed strategies, those relying on the exploitation of the intrinsic wide-band nature of *GPR* measurements appeared as effective solutions to the *ill-posedness/non-linearity* of the subsurface *IS* problem, where strongly aspect-limited set-ups are typically involved. Several techniques have been proposed to process such multi-chromatic measurements and take advantage of frequency diversity. On the one hand, frequency-hopping (*FH*) schemes [5],[10] process each frequency sample separately and exploit *progressively acquired* information at the low-pass guesses to refine the solution at successive high-frequency inversion stages. On the other hand, multi-frequency (*MF*) methods [6],[8] allow the simultaneous processing of all spectral components in a single-step reconstruction.

Besides the above-mentioned strategies, another effective recipe to counteract the *non-uniqueness* and the *ill-posedness* consists in injecting *a-priori* information about the imaged scenario. This can be done, for instance, by introducing linear models such as those based on the Born approximation in case of weak targets [9]. Otherwise, *sparseness-regularized* solvers such as those based on the Compressive Sensing (*CS*) theory [11]-[14] are able to stabilize the *GPR-IS* problem once a suitable expansion basis is chosen to represent the unknowns with few non-null coefficients.

Following this line of reasoning, this work presents three different strategies based on the integration of Bayesian *CS* (*BCS*) solution approaches with *FH* or *MF* schemes in order to benefit both from the regularization features of sparse priors as well as from multi-chromatic acquisitions.

2. Contrast source formulation of the *GPR-IS* problem

Let us consider the *2D* transverse-magnetic half-space *GPR* prospecting set-up in Fig. 1. A linear interface located at $y=0\,[\mathrm{m}]$ separates the upper half-space made of air from the lower half-space made of a lossy material with relative permittivity ε_{rB} and conductivity σ_B.

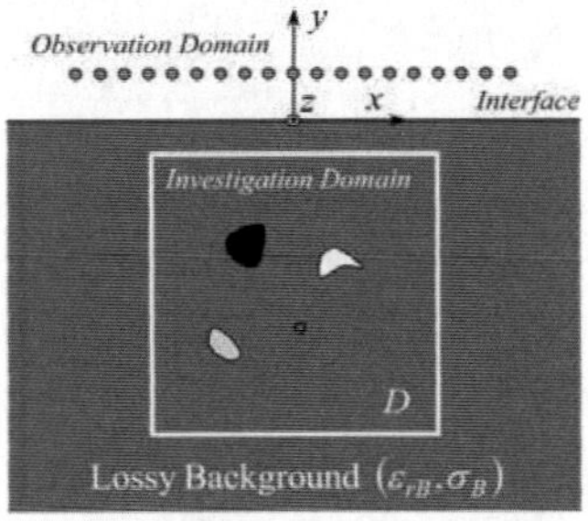

Figure 1. Geometry of the *2D GPR-IS* problem.

A square investigation domain *D*, completely buried below the interface (Fig. 1), is successively imaged by a set of *V* *z*-oriented infinitely-long line sources uniformly spaced at constant height $y=d$ and excited by a wide-band signal covering the frequency range $[f_{\min}, f_{\max}]$. Under the hypothesis that one or multiple unknown objects are embedded inside *D*, the total electric field measured in time-domain by *M* probes co-located with the sources is given by

$$e_{tot}^{v}\left(\mathbf{r}_m^v, t\right) = e_{inc}^{v}\left(\mathbf{r}_m^v, t\right) + e_{scat}^{v}\left(\mathbf{r}_m^v, t\right) + h^{v}\left(\mathbf{r}_m^v, t\right); \quad v=1,\dots,V\ ;\ m=1,\dots,M \tag{1}$$

where $e_{inc}^{v}\left(\mathbf{r}_m^v, t\right)$ is the incident field, $e_{scat}^{v}\left(\mathbf{r}_m^v, t\right)$ the scattered field, and $h^{v}\left(\mathbf{r}_m^v, t\right)$ an additive Gaussian noise super-imposed to the measured *GPR radargram*. After subtracting the incident field [8], *Q* frequency samples uniformly distributed within the $[f_{\min}, f_{\max}]$ spectrum are extracted from the (noisy) Fourier-transform of the scattered field, i.e.,

$$\hat{E}_{scat}^{v}\left(\mathbf{r}_m^v, f_q\right) = E_{scat}^{v}\left(\mathbf{r}_m^v, f_q\right) + H^{v}\left(\mathbf{r}_m^v, f_q\right); \quad v=1,\dots,V\ ;\ m=1,\dots,M\ ;\ q=1,\dots,Q\,. \tag{2}$$

$H^{v}\left(\mathbf{r}_m^v, f_q\right)$ being the frequency-domain noise. Assuming a contrast source inversion (*CSI*) scheme [15], the goal of the *GPR-IS* problem is to retrieve the *contrast source* defined for $\mathbf{r}\in D$ as follows

$$J^{v}\left(\mathbf{r}, f_q\right) = \tau\left(\mathbf{r}, f_q\right) E_{tot}^{v}\left(\mathbf{r}, f_q\right); \quad v=1,\dots,V\ ;\ q=1,\dots,Q \tag{3}$$

where $E_{tot}^{v}\left(\mathbf{r}, f_q\right)$ is the frequency-domain total field in *D*, while $\tau\left(\mathbf{r}, f_q\right)$ is the contrast function

$$\tau\left(\mathbf{r}, f_q\right) = \left[\varepsilon_r(\mathbf{r}) - \varepsilon_{rB}\right] + j\left[\frac{\sigma_B - \sigma(\mathbf{r})}{2\pi f_q \varepsilon_0}\right]; \quad q=1,\dots,Q \tag{4}$$

mathematically modelling the objects in *D* through their relative permittivity, $\varepsilon_r(\mathbf{r})$, and conductivity $\sigma(\mathbf{r})$ distributions. To numerically solve the *GPR-IS* problem, *D* is partitioned into *N* square sub-domains, yielding the following matrix expression where the real ($\Re$) and imaginary ($\Im$) parts of both data and unknowns have been separated [10]

$$\begin{bmatrix} \Re\left(\hat{\mathbf{E}}_{scat}^{v,q}\right) \\ \Im\left(\hat{\mathbf{E}}_{scat}^{v,q}\right) \end{bmatrix} = \begin{bmatrix} \Re\left(\mathbf{G}_{ext}^{v,q}\right) & -\Im\left(\mathbf{G}_{ext}^{v,q}\right) \\ \Im\left(\mathbf{G}_{ext}^{v,q}\right) & \Re\left(\mathbf{G}_{ext}^{v,q}\right) \end{bmatrix} \begin{bmatrix} \Re\left(\mathbf{J}^{v,q}\right) \\ \Im\left(\mathbf{J}^{v,q}\right) \end{bmatrix} + \begin{bmatrix} \Re\left(\mathbf{H}^{v,q}\right) \\ \Im\left(\mathbf{H}^{v,q}\right) \end{bmatrix}; \; v=1,\dots,V \; ; \quad q=1,\dots,Q . \tag{5}$$

In (5) $\hat{\mathbf{E}}_{scat}^{v,q} = \left\{\hat{E}_{scat}^{v}\left(\mathbf{r}_m^v, f_q\right); m=1,\dots,M\right\}$ are the scattered field samples under the v-th illumination for the q-th frequency, $\mathbf{J}^{v,q} = \left\{J^v\left(\mathbf{r}_n, f_q\right); n=1,\dots,N\right\}$ contains the unknown contrast sources, $H^{v,q} = \left\{H^v\left(\mathbf{r}_m^v, f_q\right); m=1,\dots,M\right\}$ contains the noise samples, while $\mathbf{G}_{ext}^{v,q}$ is the external Green's matrix for the *2D* subsurface problem. Finally, (5) is rewritten in compact form as follows

$$\boldsymbol{\xi}_{scat}^{v,q} = \boldsymbol{\Phi}_{ext}^{v,q} \boldsymbol{\psi}^{v,q} + \boldsymbol{\rho}^{v,q}; \; v=1,\dots,V \; ; \quad q=1,\dots,Q \tag{6}$$

where $\boldsymbol{\xi}_{scat}^{v,q}$ is the known data, $\boldsymbol{\Phi}_{ext}^{v,q}$ is the measurement matrix, $\boldsymbol{\psi}^{v,q}$ is the unknown, and $\boldsymbol{\rho}^{v,q}$ contains the noise samples.

3. BCS solution approaches

3.1. Frequency-hopping (FH) GPR data processing

When considering a *FH* inversion scheme, the solution of (6) is obtained by processing one frequency component of the *GPR* spectrum at a time in a cascaded fashion, starting from the lowest one. At each iteration ($q=1,\dots,Q$), the q-th unknown vector is estimated as

$$\tilde{\boldsymbol{\psi}}^{v,q} = \arg\left\{\max P\left(\boldsymbol{\psi}^{v,q} \middle| \boldsymbol{\xi}_{scat}^{v,q}\right)\right\}; \; v=1,\dots,V \tag{7}$$

where $P\left(\boldsymbol{\psi}^{v,q} \middle| \boldsymbol{\xi}_{scat}^{v,q}\right)$ is the *a-posteriori* probability. Whether the correlation among the several views ($v=1,\dots,V$) is enforced, a frequency hopping multi-task *BCS* (*FH-MT-BCS*) solution strategy can adopted, whose solution is given by [16]

$$\tilde{\boldsymbol{\psi}}_{FH-MT-BCS}^{v,q} = \left[diag\left(\tilde{\boldsymbol{\alpha}}^q\right) + \left(\boldsymbol{\Phi}_{ext}^{v,q}\right)^T \boldsymbol{\Phi}_{ext}^{v,q} \right]^{-1} \left(\boldsymbol{\Phi}_{ext}^{v,q}\right)^T \boldsymbol{\xi}_{scat}^{v,q} \; ; \; v=1,\dots,V \; . \tag{8}$$

In (8) $.^T$ is the transpose operator, while $\tilde{\boldsymbol{\alpha}}^q$ is the set of $2N$ hyper-parameters shared among the V views, determined by solving through a constrained relevance vector machine (*C-RVM*) solver [10]

$$\tilde{\boldsymbol{\alpha}}^q = \arg\left\{ \max_{\boldsymbol{\alpha}^q} \left(-\frac{1}{2} \sum_{v=1}^{V} \left[\left(2M + 2\beta_1\right) \log\left(\left(\boldsymbol{\xi}_{scat}^{v,q}\right)^T \left(\mathbf{B}^{v,q}\right)^{-1} \boldsymbol{\xi}_{scat}^{v,q} + 2\beta_2 \right) + \log\left|\mathbf{B}^{v,q}\right| \right] \right) \right\} \tag{9}$$

where $\mathbf{B}^{v,q} = \mathbf{I} + \boldsymbol{\Phi}_{ext}^{v,q} \left[diag\left(\boldsymbol{\alpha}^q\right)\right]^{-1} \left(\boldsymbol{\Phi}_{ext}^{v,q}\right)^T$ and $\left(\beta_1, \beta_2\right)$ are user-defined *MT-BCS* parameters [16] and $\mathbf{I}$ the identity matrix. Given the fact that a pixel basis is exploited to expand the unknown, each hyper-parameter will physically correspond to the real or imaginary part of the contrast source inside the N pixels within D.

On the other hand, if no correlation is enforced among the different illuminations a single-task procedure (*FH-ST-BCS*) can be adopted for solving (7), the contrast sources being estimated as

$$\tilde{\boldsymbol{\psi}}_{FH-ST-BCS}^{v,q} = \frac{1}{\tilde{\sigma}^{v,q}} \left[\frac{\left(\boldsymbol{\Phi}_{ext}^{v,q}\right)^T \boldsymbol{\Phi}_{ext}^{v,q}}{\tilde{\sigma}^{v,q}} + diag\left(\tilde{\boldsymbol{\delta}}^{v,q}\right) \right]^{-1} \left(\boldsymbol{\Phi}_{ext}^{v,q}\right)^T \boldsymbol{\xi}_{scat}^{v,q} \; ; \; v=1,\dots,V \; . \tag{10}$$

where $\tilde{\sigma}^{v,q}$ and $\tilde{\boldsymbol{\delta}}^{v,q}$ are the estimated noise variance and the auxiliary *ST-BCS* hyper-parameters (different among the views, $v=1,\dots,V$), computed via the *C-RVM* solver as

$$\left(\tilde{\boldsymbol{\delta}}^{v,q},\tilde{\sigma}^{v,q}\right)=\arg\left\{\max_{\left(\boldsymbol{\delta}^{v,q},\boldsymbol{\sigma}^{v,q}\right)}\left(-\frac{1}{2}\left[2M\log 2\pi+\log\left|\mathbf{W}^{v,q}\right|+\left(\boldsymbol{\xi}_{scat}^{v,q}\right)^{T}\left(\mathbf{W}^{v,q}\right)^{-1}\boldsymbol{\xi}_{scat}^{v,q}\right]\right)\right\} \tag{11}$$

where $\mathbf{W}^{v,q}=\sigma^{v,q}\mathbf{I}+\boldsymbol{\Phi}_{ext}^{v,q}\left[diag\left(\boldsymbol{\delta}^{v,q}\right)\right]^{-1}\left(\boldsymbol{\Phi}_{ext}^{v,q}\right)^{T}$. Both the *FH-MT-BCS* and the *FH-ST-BCS* exploit the *acquired* information about the solution at the q-th frequency to initialize the *C-RVM* optimization at the $(q+1)$-th step. Towards this end, at each q-th step ($q>1$) the set of hyper-parameters to optimize is progressively shrunk to those entries corresponding to cells in D falling inside the identified regions-of-interest (i.e., sub-regions of D where a potential scatterer has been located at the q-th inversion stage) [10]. Finally, the estimated contrast $\tilde{\tau}\left(\mathbf{r}_n, f_q\right)$, $n=1,\dots,N$, is computed at each frequency from (3) by averaging over the different views ($v=1,\dots,V$) the ratio between retrieved currents and the corresponding retrieved total fields.

3.2. Multi-frequency (MF) GPR data processing

In order to exploit the frequency diversity of wide-band *GPR* data in a single reconstruction a multi-frequency multi-task *BCS* (*MF-MT-BCS*) approach can be adopted. In this case, the unknown contrast sources are estimated by imposing in (7) that $\boldsymbol{\psi}^{v,q}$ are correlated among the different frequency samples ($q=1,\dots,Q$) other than among the several views ($v=1,\dots,V$). Accordingly, the *MF-MT-BCS* solution is computed as

$$\tilde{\boldsymbol{\Psi}}_{MF-MT-BCS}^{v,q}=\left[diag\left(\tilde{\boldsymbol{\upsilon}}\right)+\left(\boldsymbol{\Phi}_{ext}^{v,q}\right)^{T}\boldsymbol{\Phi}_{ext}^{v,q}\right]^{-1}\left(\boldsymbol{\Phi}_{ext}^{v,q}\right)^{T}\boldsymbol{\xi}_{scat}^{v,q}\ ;\ v=1,\dots,V\ ;\ q=1,\dots,Q \tag{12}$$

where the $2N$ shared hyper-parameters $\tilde{\boldsymbol{\upsilon}}$ (among all views and frequencies) are estimated via a *RVM* solver [16] as

$$\tilde{\boldsymbol{\upsilon}}=\arg\left\{\max_{\boldsymbol{\upsilon}}\left(-\frac{1}{2}\sum_{q=1}^{Q}\sum_{v=1}^{V}\left[\left(2M+2\beta_1\right)\log\left(\left(\boldsymbol{\xi}_{scat}^{v,q}\right)^{T}\left(\mathbf{R}^{v,q}\right)^{-1}\boldsymbol{\xi}_{scat}^{v,q}+2\beta_2\right)+\log\left|\mathbf{R}^{v,q}\right|\right]\right)\right\}. \tag{13}$$

where $\mathbf{R}^{v,q}=\mathbf{I}+\boldsymbol{\Phi}_{ext}^{v,q}\left[diag\left(\boldsymbol{\upsilon}\right)\right]^{-1}\left(\boldsymbol{\Phi}_{ext}^{v,q}\right)^{T}$. The contrast function in D is then estimated by averaging over the different frequencies ($q=1,\dots,Q$) and views ($v=1,\dots,V$) the ratio between retrieved currents and the corresponding retrieved total fields.

4. Numerical comparisons

In order to comparatively assess the proposed *BCS*-based inversion methodologies for solving the *GPR-IS* problem let us consider in the following an investigation domain of side 0.8m buried in a lossy soil with $\varepsilon_{rB}=4$ and $\sigma_B=10^{-3}$ S/m. $V=20$ sources placed at $d=0.1$m over the interface are used to successively probe D with wide-band pulses in the $\left[f_{\min}, f_{\max}\right]=\left[200,600\right]$MHz band. $M=19$ probes are used each time to collect the *GPR* time-domain data, which is successively transformed to the frequency-domain where $Q=9$ spectral components are extracted for the inversion. The *FDTD*-based solver *GPRMax2D* [8] has been exploited to synthesize the time-domain radargrams, while a custom FORTRAN implementation of the *2D* Method of Moments (*MoM*) has been adopted for implementing the external Green's matrix and solving the inverse problem.

Figure 2 shows the real part of the contrast retrieved by the *MF-MT-BCS* [Fig. 2(*b*)], the *FH-MT-BCS* [Fig. 2(*c*)], and the *FH-ST-BCS* [Fig. 2(*d*)] when considering the "*Three-Pixel*" scatterer in Fig. 2(*a*) ($\varepsilon_r^{obj}=5.5$ and $\sigma^{obj}=10^{-4}$ S/m) and a signal-to-noise-ratio (on time-domain total field) equal to $SNR=40$ dB. As it can be seen, faithful reconstructions are yielded by the two *MT-BCS* approaches, which are both able to correctly detect the presence of the scatterer [Figs. 2(*b*)-2(*c*)], while the *FH-ST-BCS* solution is less accurate with a lot of undesired artifacts in the background region [Fig. 2(*d*)].

IOP Conf. Series: Journal of Physics: Conf. Series **1131** (2018) 012006 doi:10.1088/1742-6596/1131/1/012006

Moreover, it should be pointed out that imposing the correlation among different frequency components of the scattered data allows a significant enhancement of the *MF-MT-BCS* solution quality [Fig. 2(*b*)] with respect to the *FH-MT-BCS* one [Fig. 2(*c*)].

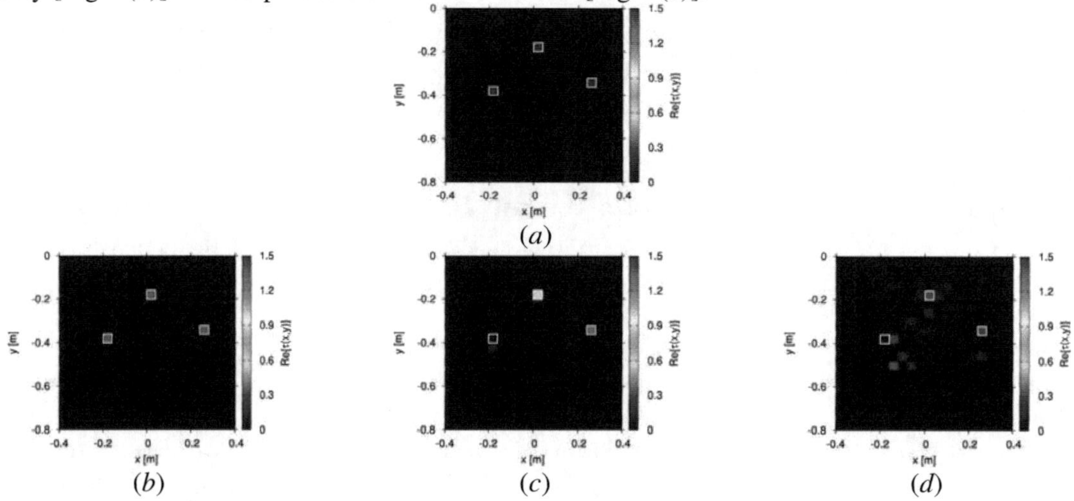

Figure 2. *Numerical Comparison (Three-Pixels Scatterer,* $\varepsilon_r^{obj}=5.5$, $\sigma^{obj}=10^{-4}$ S/m, $SNR=40\,\text{dB}$) – Real part of (*a*) the actual contrast profile and of the retrieved solutions by the (*b*) *MF-MT-BCS*, (*c*) *FH-MT-BCS,* and (*d*) *FH-ST-BCS* methods.

Similar conclusions can be drawn by looking at the imaginary part of the retrieved contrasts in Fig. 3, where the superiority of the multi-frequency inversion strategy is even more evident over the two frequency-hopping alternatives [Fig. 3(*b*) vs. Figs. 3(*c*)-3(*d*)].

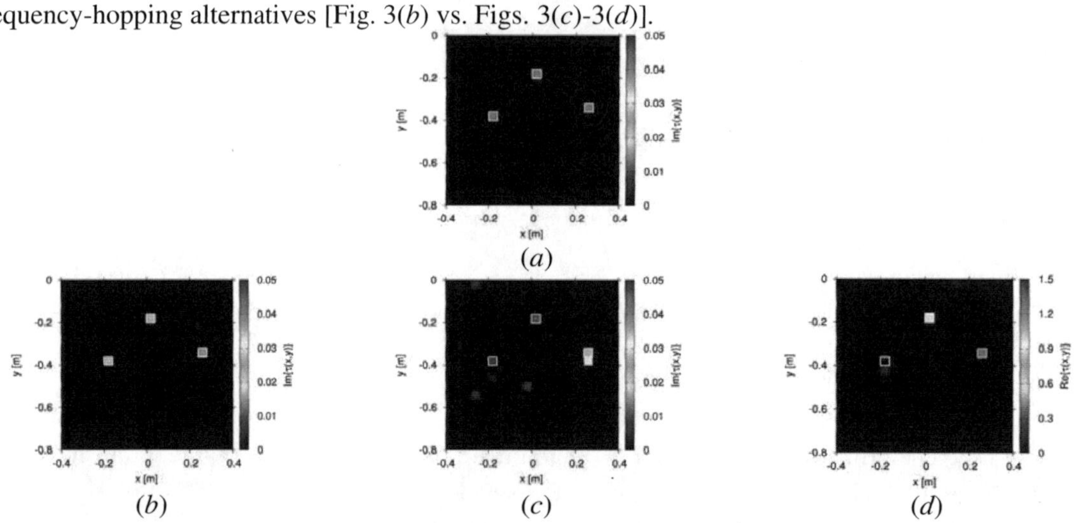

Figure 3. *Numerical Comparison (Three-Pixels Scatterer,* $\varepsilon_r^{obj}=5.5$, $\sigma^{obj}=10^{-4}$ S/m, $SNR=40\,\text{dB}$) – Imaginary part of (*a*) the actual contrast profile and of the retrieved solutions by the (*b*) *MF-MT-BCS*, (*c*) *FH-MT-BCS,* and (*d*) *FH-ST-BCS* methods.

To further investigate the performance of the three *BCS GPR-IS* methods, let us consider in the following the retrieval of the "*Two-Bars*" profile in Fig. 5(*a*) ($\varepsilon_r^{obj}=7.0$, $\sigma^{obj}=10^{-3}$ S/m). A significant boost in terms of reconstruction accuracy is achieved by the *MF-MT-BCS* [Fig. 5(*b*)], providing a better estimation of the contrast value with respect to the *FH-MT-BCS* [Fig. 5(*c*)] and a remarkably improved solution when compared to the *FH-ST-BCS* [Fig. 5(*d*)].

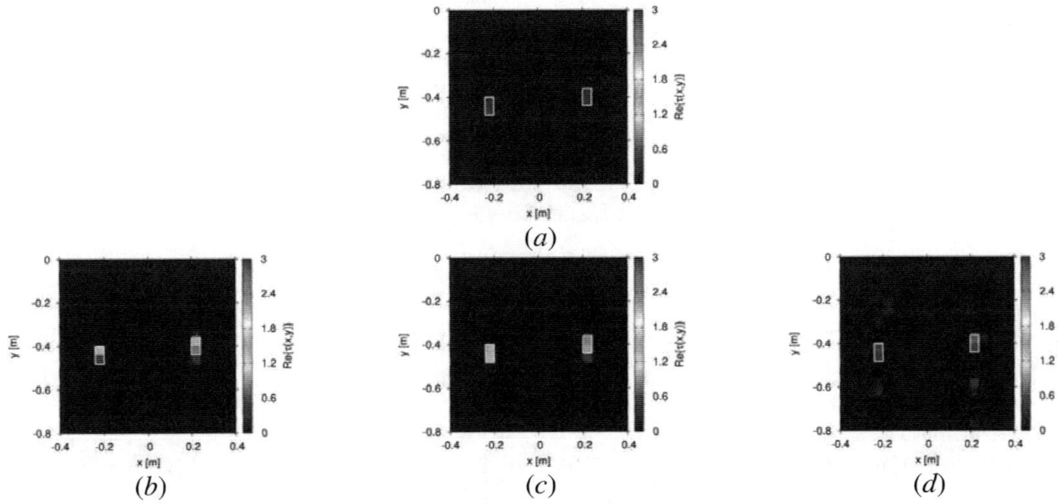

Figure 4. *Numerical Comparison* (*Two-Bars Scatterer,* $\varepsilon_r^{obj} = 7.0$, $\sigma^{obj} = 10^{-3}$ S/m, $SNR = 40$ dB) – Real part of (*a*) the actual contrast profile and of the retrieved solutions by the (*b*) *MF-MT-BCS*, (*c*) *FH-MT-BCS,* and (*d*) *FH-ST-BCS* methods.

Finally, the total reconstruction error, defined as in [10] is reported in Fig. 5 as a function of the actual relative permittivity of the same scatterer of Fig. 4(*a*). As it can be noticed, the *FH-ST-BCS* solutions are always significantly worse than those of the two *MT* approaches since no correlation among the different views is exploited. Moreover, this result further verifies that the *MF-MT-BCS* is the more robust strategy, leading to lower reconstruction errors in presence of both weak and strong targets (Fig. 5).

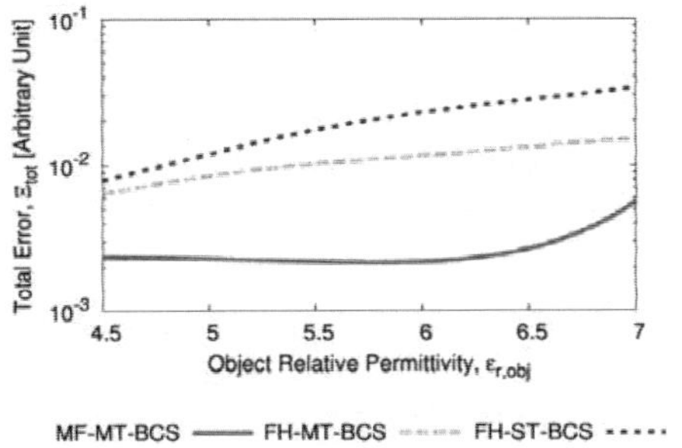

Figure 5. *Numerical Comparison* (*Two-Bars Scatterer,* $\sigma^{obj} = 10^{-3}$ S/m, $SNR = 40$ dB) – Total reconstruction error versus the relative permittivity of the object.

5. Conclusions

The processing of wide-band *GPR* data through *BCS*-based inverse scattering techniques has been addressed in this work. The proposed approaches allow to exploit the intrinsic frequency-diversity of *GPR* measurements through frequency-hopping (*FH*) or multi-frequency (*MF*) solution strategies, as well as benefit from the sparseness-driven regularization features of *BCS*. From a methodological point of view, the main novelties are the development of a *C-RVM* solver to exploit progressively acquired information from low-pass to high-pass solutions in *FH* schemes and the formulation of the *GPR-IS* problem within a *MF-CSI* framework enabling the exploitation of the existing correlation among several frequencies and views. Selected numerical results have been shown in order to comparatively assess the performances of three *BCS* implementations. The reported results indicate that (*i*) correlating the several views through multi-task (*MT*) approaches allow to significantly reduce the reconstruction error with respect to single-task (*ST*) strategies and (*ii*) jointly processing the frequency samples of the *GPR* spectrum through the *MF-MT-BCS* method provides more robust reconstructions under several operative conditions. Future works will be devoted at assessing the three techniques on real experimental data, as well as to extend them to deal with a *3D* formulation of the subsurface imaging problem.

Acknowledgment
This work has been partially supported by the Italian Ministry of Foreign Affairs and International Cooperation, Directorate General for Cultural and Economic Promotion and Innovation within the SNATCH Project (2017-2019).

References

[1] Peters L, Daniels J and Young J D 1994 Ground penetrating radar as a subsurface environmental sensing tool *IEEE Proc.* **82** 1802–1822
[2] Lesselier D and Habashy T 2000 Special issue on electromagnetic imaging and inversion on the Earth's subsurface *Inverse Probl.* **16**
[3] Li M, Abubakar A and Habashy TM 2010 Application of a two-and-half dimensional model-based algorithm to crosswell electromagnetic data inversion *Inverse Probl.* **26** 1-17
[4] Abubakar A, Habashy TM, Li M and Liu J 2009 Inversion algorithm for large-scale geophysical electromagnetic measurements *Inverse Probl.* **25** 1-30
[5] Salucci M, Oliveri G and Massa A 2015 GPR prospecting through an inverse scattering frequency-hopping multifocusing approach *IEEE Trans. Geosci. Remote. Sens.* **53** 6573-6592
[6] Salucci M, Poli L and Massa A 2017 Advanced multi-frequency GPR data processing for non linear deterministic imaging *Signal Processing* **132** 306-318
[7] Krueger KR, McClellan JH, Scott WR 2015 Efficient algorithm design for GPR imaging of landmines *IEEE Trans. Geosci. Remote Sens.* **53** 4010-4021
[8] Salucci M, Poli L, Anselmi N and Massa A 2017 Multifrequency particle swarm optimization for enhanced multiresolution GPR microwave imaging *IEEE Trans. Geosci. Remote Sens.* **55** 1305-1317
[9] Salucci M, Oliveri G, Randazzo A, Pastorino M and Massa A 2014 Electromagnetic subsurface prospecting by a multifocusing inexact Newton method within the second-order Born approximation *J. Opt. Soc. Am. A* **31** 1167-1179
[10] Salucci M, Gelmini A, Poli L, Oliveri G and Massa A 2018 Progressive compressive sensing for exploiting frequency-diversity in GPR imaging *Journal of Electromagnetic Waves and Applications* ***32*** *1164-1193*
[11] Oliveri G, Salucci M, Anselmi N and Massa A 2017 Compressive sensing as applied to inverse problems for imaging: theory, applications, current trends, and open challenges *IEEE Antennas Propag. Mag.* **59** 34-46
[12] Sun Y, Qu L, Zhang S and Yin Y 2016 MT-BCS-based two-dimensional diffraction tomographic GPR imaging algorithm with multiview-multistatic configuration *IEEE Geosci. Remote Sens. Lett.* **13** 831-835
[13] Suksmono AB, Bharata E, Lestari AA, Yarovoy AG and Lightart LP 2010 Compressive stepped-frequency continuous-wave ground-penetrating radar *IEEE Geosci. Remote Sens. Lett.* **7** 665-669
[14] Anselmi N, Oliveri G, Hannan M A, Salucci M and Massa A 2017 Color compressive sensing imaging of arbitrary-shaped scatterers *IEEE Trans. Microw. Theory Techn.* **65** 1986-1999
[15] Li M, Semerci O and Abubakar A 2013 A contrast source inversion method in the wavelet domain *Inverse Probl.* **29** 025015
[16] Ji S, Dunson D and Carin L 2009 Multitask compressive sensing *IEEE Trans. Signal Process.* **57** 92-106

Heuristics For Efficient Sparse Blind Source Separation

C Kervazo[1], J Bobin[1] and C Chenot[2]

[1] CEA Saclay, Gif-sur-Yvette, 91191 cedex, France
[2] Institute for Digital Communications, University of Edinburgh, UK

E-mail: christophe.kervazo@cea.fr

Abstract. Sparse Blind Source Separation (sparse BSS) is a key method to analyze multichannel data in fields ranging from medical imaging to astrophysics. However, since it relies on seeking the solution of a *non-convex* penalized matrix factorization problem, its performances largely depend on the optimization strategy. In this context, Proximal Alternating Linearized Minimization (PALM) has become a standard algorithm which, despite its theoretical grounding, generally provides poor practical separation results. In this work, we first investigate the origins of these limitations, which are shown to take their roots in the sensitivity to both the initialization and the regularization parameter choice. As an alternative, we propose a novel strategy that combines a heuristic approach with PALM. We show its relevance on realistic astrophysical data.

1. Introduction

1.1. Blind Source Separation problem

In the BSS [1] framework, the data are composed of m observations, each of which has t samples. These observations are supposed to be some linear combinations of n sources. The objective of BSS is to retrieve the sources as well as the mixing coefficients. In matrix form, the goal is therefore to find two matrices $\mathbf{S}$ (of size $n \times t$) and $\mathbf{A}$ (of size $m \times n$), called respectively the source and the mixing matrices, such that: $\mathbf{X} = \mathbf{AS} + \mathbf{N}$, where $\mathbf{X}$ (of size $m \times t$) is the observation matrix that is corrupted with some unkwown noise $\mathbf{N}$. Since it requires tackling an ill-posed unsupervised matrix factorization problem, further assumptions are needed, including the statistical independance of the sources (ICA - [1]), the non-negativity of $\mathbf{A}$ and $\mathbf{S}$ [2]. In this work, we will focus on the sparsity of the sources [3–6]. In this framework, sparse BSS will aim at finding a (local) minimum of:

$$\min_{\mathbf{A},\mathbf{S}} \frac{1}{2}\|\mathbf{X} - \mathbf{AS}\|_F^2 + \iota_{\{\forall i; \|\mathbf{A}^i\|_2^2 \leq 1\}}(\mathbf{A}) + \left\|\mathbf{R_S} \odot (\mathbf{S}\mathbf{\Phi}_\mathbf{S}^\mathbf{T})\right\|_1 \quad (1)$$

The first data fidelity term promotes a faithful reconstruction of the data. The use of the Froebenius norm $\|.\|_F$ stems from the assumption of a Gaussian noise $\mathbf{N}$. The second term corresponds to the oblique constraint ensuring that the columns of $\mathbf{A}$ are all in the ℓ_2 ball. This avoids degenerated $\mathbf{A}$ and $\mathbf{S}$ matrices where $\|\mathbf{A}\| \to \infty$ and $\|\mathbf{S}\| \to 0$. The last term involving the Hadamard product $\odot$ enforces a ℓ_1 sparsity constraint in a transformed domain $\mathbf{\Phi_S}$. In the following, $\mathbf{\Phi_S}$ will be taken equal to either the identity (in which case the sparsity is enforced in the direct domain) or the starlet transform. $\mathbf{R_S}$ controls the trade-off between the data fidelity and the sparsity terms. It can be decomposed into $\mathbf{R_S} = \mathbf{\Lambda_S} \odot \mathbf{W}$ where $\mathbf{\Lambda_S}$ is a diagonal matrix of the regularization parameters $\lambda_1, \lambda_2, ..., \lambda_n$ and $\mathbf{W}$ is a matrix used to introduce individual penalization coefficients in the context of reweighted ℓ_1 [7] (when no reweighting is used, $\mathbf{W}$ is equal to the identity matrix).

1.2. Sparse BSS in practice
Since sparse BSS requires solving a penalized matrix factorization problem, it is important to highlight that the separation quality strongly depends on the optimization strategy. For that purpose, different algorithmic frameworks have been used so far: projected Alternate Least-Square (ALS - [2]), Block-Coordinate Descent (BCD - [8]) and PALM [9]. However, any practitioner can draw the same conclusion: the solution of sparse BSS methods is highly sensitive to the initial point and the values of the regularization parameters, which are generally tricky to tune without any first guess of the solution.

- *Initialization:* as problem (1) is not convex but multi-convex, the algorithm performing its minimization can be trapped in spurious critical points, depending on the initial matrices $\mathbf{A}$ and $\mathbf{S}$. Different optimization strategies can be affected differently.
- *Regularization parameters:* the practical choice of the regularization parameters $\mathbf{\Lambda_S}$ is of paramount significance:
  - The parameters are directly impacting the shape of the solution through the trade-off between the sparsity level and fidelity to the data.
  - The problem being non-convex, a change in the parameters can also bring the optimization algorithm to stabilize in the neighborhood of a different critical point, potentially leading to a very different solution.

1.3. Contributions
While PALM is theoretically well rooted and yields rather fast minimization schemes (in contrast to BCD), it generally provides poor separation results. We first investigate the origins of this poor practical efficiency, and especially the sensitivity to both initialization and regularization parameter settings. We further show how PALM-based implementations can benefit from the information provided by heuristic approaches which are in pratice more robust. The robustness of the proposed combined strategy is demonstrated on realistic astrophysical data.

2. Optimization strategy: PALM and GMCA
As emphasized before, the optimization strategy is decisive to avoid stationary points. In the following, we will discuss PALM [9] and projected ALS such as Generalized Morphological Component Analysis (GMCA - [3]). BCD will not be studied since it requires an exact minimization at each iteration, leading in general to a high computational cost.

2.1. PALM algorithm
PALM [9] attractiveness partially comes from the fact that it is proved to converge to a local minimum of (1) under mild conditions [9]. PALM is an iterative algorithm. At each iteration, it alternates between a proximal gradient step on $\mathbf{A}$ and $\mathbf{S}$. In our case, it can be shown that the proximal operator corresponding to the ℓ_1 sparsity term in (1) is the soft-thresholding operator $\mathcal{S}_{\Lambda_\mathbf{S}}(.)$ applied in the transform domain. Furthermore, the operator corresponding to the oblique constraint is the projection of each column of $\mathbf{A}$ on the ℓ_2 unit ball, which we shall denote as $\Pi_{\|.\|_2}(.)$. Each iteration (k) can then be read as:

1 - Update of $\mathbf{S}$ using a fixed $\mathbf{A}$:

$$\hat{\mathbf{S}} = \mathbf{S}^{(k-1)} - \frac{\gamma}{\left\|\mathbf{A}^{(k-1)T}\mathbf{A}^{(k-1)}\right\|_2}\mathbf{A}^{(k-1)T}(\mathbf{A}^{(k-1)}\mathbf{S}^{(k-1)} - \mathbf{X}) \quad (2)$$

$$\mathbf{S}^{(k)} = \mathcal{S}_{\frac{\gamma \mathbf{R_S}}{\left\|\mathbf{A}^{(k-1)T}\mathbf{A}^{(k-1)}\right\|_2}}\left(\hat{\mathbf{S}}\mathbf{\Phi}_\mathbf{S}^\mathbf{T}\right)\mathbf{\Phi}_\mathbf{S} \quad (3)$$

2 - Update of $\mathbf{A}$ using a fixed $\mathbf{S}$:

$$\hat{\mathbf{A}} = \mathbf{A}^{(k-1)} - \frac{\delta}{\left\|\mathbf{S}^{(k)}\mathbf{S}^{(k)T}\right\|_2}(\mathbf{A}^{(k-1)}\mathbf{S}^{(k)} - \mathbf{X})\mathbf{S}^{(k)T} \quad (4)$$

$$\mathbf{A}^{(k)} = \Pi_{\|.\|_2}\left(\hat{\mathbf{A}}\right) \quad (5)$$

2.2. GMCA algorithm
GMCA algorithm is based on projected ALS. At each iteration (k), the gradient step of (2) and (4) is replaced by a multiplication by a pseudo-inverse: $\hat{\mathbf{S}} = \mathbf{A}^{(k-1)\dagger}\mathbf{X}$ and $\hat{\mathbf{A}} = \mathbf{X}\mathbf{S}^{(k)\dagger}$. Compared to PALM, GMCA cannot be proved to converge (while stabilizing in most of practical cases after some iterations). Furthermore, even when stabilizing the output of GMCA is not exactly minimizing (1) but rather an approximation of the cost function. However, thanks to heuristics, it benefits from an automatic thresholding strategy (see Sec.4.1.1) which has been empirically shown to improve its robustness with respect to the initialization.

3. PALM-based sparse BSS in practice
The objective of this section is to empirically shed light on the lack of robustness of PALM to motivate the need of finding a relevant strategy to enable its use on practical problems.

3.1. Experimental protocol
To bring out PALM sensibility and the mechanisms at stake, our experiments focus in this subsection on simple data coming from $n = 2$ equilibrated (with equal ℓ_2 norm) sources. The sources are assumed to be exactly sparse in the direct domain, a proportion $p = 0.1$ of the $t = 500$ samples being non-zeros and drawn according to a standard normal distribution. The mixing is performed through a $\mathbf{A}$ matrix drawn randomly following a standard normal distribution and modified to have unit columns. Its condition number is $C_d = 10$. There is $m = n$ observations. To complete the creation of the $\mathbf{X}$ data, a Gaussian additive noise is added to the mixing, such that the Signal to Noise Ratio is SNR = 60 dB.
Once the algorithm launched, it stops based on a convergence criterion computed through the angular distance of the columns of $\mathbf{A}$ between two successive iterations. The separation quality is then mesured using a global criterion on $\mathbf{A}$ [10]: $C_A = -10\log_{10}(\texttt{median}(|\mathbf{PA}^\dagger\mathbf{A}^*| - \mathbf{I_d}))$, with $\mathbf{A}^*$ the true mixing matrix and $\mathbf{PA}^\dagger$ the pseudo-inverse of the solution found by the algorithm corrected through $\mathbf{P}$ for the scale and permutation indeterminacies. The higher C_A, the better the separation.
To study PALM robustness, we perform an exhaustive search for the parameters. This would be intractable on practical cases with a higher number of sources and unknown true $\mathbf{A}^*$ and $\mathbf{S}^*$. In this section, the parameters are held fixed during the whole algorithm, the goal being to show the sensibility to the parameter choice. The results might however be sub-optimal due to the fact that varying parameters could define a better trajectory avoiding spurious critical points.

3.2. Sensibility to the thresholding strategy
Following our experimental protocol, we obtain Figure 1, on which the mean of C_A is displayed as a function of the threshold of each of the 2 sources. The lack of robustness and sensibility to the parameter choice is underlined by the following remarks:

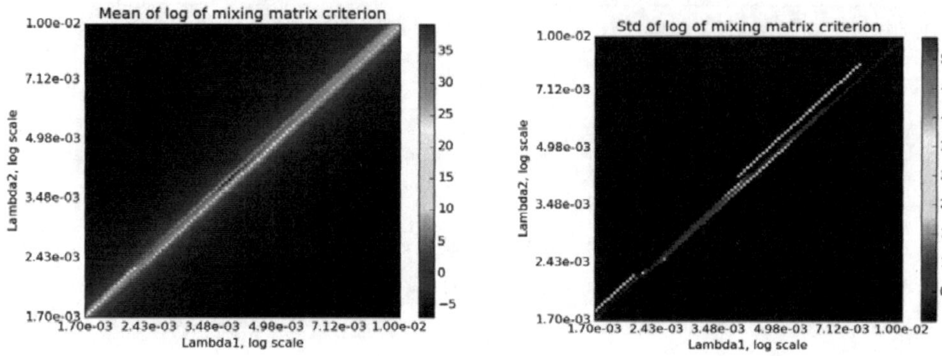

Figure 1. *Left:* Mean of C_A as a 2D function of the thresholds for 5 different initializations of PALM algorithm. *Right:* Standard deviation of C_A as a 2D function of the thresholds for 5 different initializations of PALM algorithm.

- High quality separation happens for values close to the diagonal. A slight deviation from it can cause a 30 dB drop for a 3×10^{-4} modification of one parameter and lead to a very bad separation. One could argue that this problem could be circumvented by imposing $\lambda_1 = \lambda_2$. First, it has to be emphasized that the maximum is *not* on the diagonal but slightly outside, for $\lambda_1 = 3.81 \times 10^{-3}$ and $\lambda_2 = 3.88 \times 10^{-3}$. Merely setting $\lambda_1 = \lambda_2 = 3.81 \times 10^{-3}$ yields a loss of more than 10 dB. Second, in this simple toy example, the sources are equilibrated. It would not necessarily be the case with real-life sources, which could lead to a change in the diagonal structure of the figure;
- Inside of the diagonal itself, a 2×10^{-3} change of $\lambda_1 = \lambda_2$ can lead to a loss of about 7 dB.

3.3. Sensibility to the initialization
As described in section 1.2, the initialization impacts the results due to the non-convex nature of problem (1) because the algorithm can stay trapped in spurious critical points. The standard deviation of C_A over different initializations of $\mathbf{A}$ is plotted in Fig. 1 to quantify this impact. It can reach up to 5 dB for some regularization parameters choices. The best parameter setting corresponds to a quite high standard deviation and can shift depending on the initialization. It makes the practical choice of an initialization a critical point which directly impact the quality of the local minimum found by the algorithm.
It further highlights the sensitivity to the values of the regularization parameters. A key question is thus *how to tune PALM in practice and if it can be done in an automatic way without prior knowledge about the solution.*

4. Complexity of introducing heuristics in PALM

4.1. Heuristic motivation and description
Building on the automatic thresholding strategy of GMCA, the goal of this part is to try to derive one for PALM.

4.1.1. Automatic parameter choice in GMCA In GMCA, the threshold choice is performed computing the Median Absolute Deviation (MAD) for each source and setting the corresponding threshold to a multiple k of this value:

$$(\lambda_1, \lambda_2, ..., \lambda_n)^T = k \times \texttt{MAD}(\hat{\mathbf{S}}) \tag{6}$$

In this equation and in the remaining of this section, $\mathbf{\Phi}_\mathbf{S}^\mathbf{T}$ was supposed to be the identity matrix without loss of generality. k is a positive number and the MAD operator is computed row-by-row with $\texttt{MAD}(z) = \texttt{median}(|z - \texttt{median}(z)|)$ for $z \in \mathbb{R}^t$. If we suppose that GMCA has found the true $\mathbf{A}$ matrix denoted by $\mathbf{A}^*$, it can be shown that:

$$k \times \texttt{MAD}(\hat{\mathbf{S}}) \simeq k \times \texttt{MAD}\left(\mathbf{A}^{*\dagger}\mathbf{N}\right) \tag{7}$$

The true thresholds should therefore remove a dense signal coming from the back-projection of the data noise $\mathbf{N}$. This interpretation of the thresholding is however limited to the case when the estimation of $\mathbf{A}$ is close to $\mathbf{A}^*$. In the opposite case, the unperfect demixing results in an increase of the thresholds due to an extra noise coming from interferences between the sources.

4.1.2. Heuristic in PALM Lets assume that the thresholds in PALM are computed the same way as in GMCA through eq. (6), and that the algorithm has converged to *both* the true matrices $\mathbf{A}^*$ and $\mathbf{S}^*$. Then, due to the assumption of $\mathbf{S}^*$ sparsity:

$$(\lambda_1, \lambda_2, ..., \lambda_n)^T \simeq k \times \frac{\gamma}{\|\mathbf{A}^{*T}\mathbf{A}^*\|_2}\texttt{MAD}(\mathbf{A}^{*T}\mathbf{N}) \tag{8}$$

Therefore, using the MAD enables a thresholding of a projection of the noise $\mathbf{N}$, which yields a similar interpretation as in GMCA. This observation must however be tempered:

- It only holds when and if PALM has converged towards good $\mathbf{A}$ *and* $\mathbf{S}$;
- The projection is not performed in the same space as GMCA, for which the noise is back-projected in the *source* space. Both spaces however merge when $\mathbf{A}$ is orthogonal.

Finally, it has to be noted that the thresholds become fixed after a given number of iterations to ensure the convergence.

4.2. Experimental protocol

The mixing and the metric on the separation quality are the same as in sec. 3.1. The automatic parameter setting described in the previous subsection is used inside of a PALM algorithm for different k values.

4.3. Empirical performances of the heuristic

The results are displayed in Fig 2. The low values of C_A, reaching at most 6dB, are to be compared with the 39 dB obtained during the exhaustive parameter search in 3.1. While efficient in GMCA, the MAD heuristic leads to a bad separation quality due to the different space in which the thresholding is performed (cf. 4.1.2), yielding a different role to the thresholds. To better understand this role, lets assume that the algorithm is initialized with the true mixing matrix $\mathbf{A}^*$ *only* (and not with $\mathbf{S}^*$, contrary to what is described in 4.1.2). In PALM, the use of the transpose $\mathbf{A}^{*T}$ in eq. (2) leads towards a re-mixing of the sources at each iteration whose goal is to gradually decrease the data fidelity term. This re-mixing however increases the sparsity term and should be impeded by the thresholds, giving to them a demixing role.
On the opposite, when using GMCA, such a re-mixing between the sources does not exist due to the use of the pseudo-inverse $\mathbf{A}^{*\dagger}$ instead of $\mathbf{A}^{*T}$, which alleviate somehow the issue of finding relevant thresholds. Therefore, using the MAD heuristic in PALM does not results in good separations in practice, because the interpretation in terms of noise removal is less relevant than in GMCA.

IOP Conf. Series: Journal of Physics: Conf. Series **1131** (2018) 012007 doi:10.1088/1742-6596/1131/1/012007

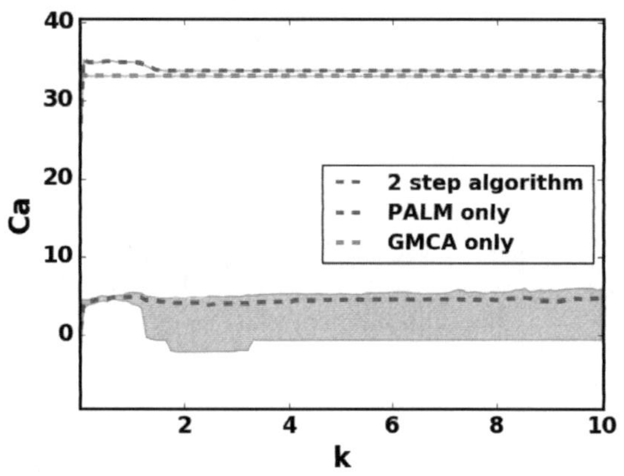

Figure 2. Mean and quartiles of C_A for 10 different initializations.

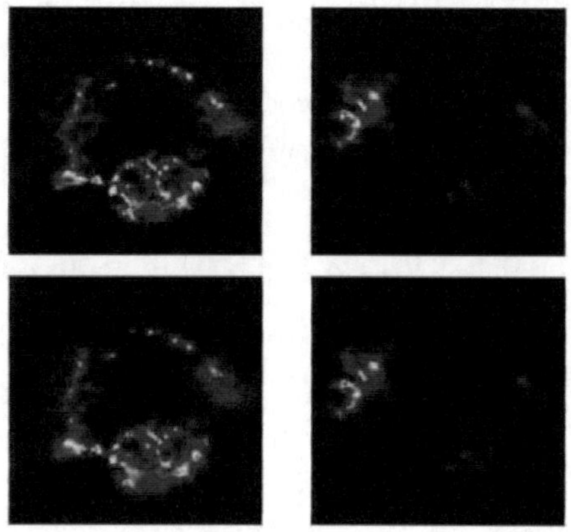

Figure 3. *Up:* true sources. *Down:* sources estimated by the two step algorithm.

5. Combining GMCA and PALM: a hybrid strategy

In this part, we propose to combine the best of GMCA and PALM in a two step approach. The algorithm comprehends a warm-up stage, in which GMCA is performed, followed by a refinement stage during which PALM is performed retaining as much information as possible coming from the warm-up stage.

5.1. Motivation and full description of the algorithm

Our approach is motivated by several remarks:

- *PALM theoretical background:* in particular, the 2-step algorithm is thus proved to converge.
- *GMCA robustness with regards to initialization*, which will lead to a robust 2-step algorithm.
- *Benefit from GMCA solution:*
  - Since *both* $\mathbf{A_{GMCA}}$ and $\mathbf{S_{GMCA}}$ are close to $\mathbf{A}^*$ and $\mathbf{S}^*$, they can be used to derive the thresholds using the MAD (after a given number of iterations the thresholds become fixed to ensure convergence).
  - $\mathbf{S_{GMCA}}$ should already give a good approximation of the most prominent peaks. This information can be exploited in the refinement stage through the introduction of reweighted L1 [7] using the reweighting matrix $\mathbf{W}$ in problem (1).

These remarks lead to the following 2-step algorithm:

Input : $\mathbf{X}$ (data matrix)

- Random initialization $\mathbf{A_0}$ and $\mathbf{S_0}$
- *Warm-up stage*:
  $\mathbf{A_{GMCA}}$, $\mathbf{S_{GMCA}}$ = GMCA($\mathbf{X}$,$\mathbf{A_0}$,$\mathbf{S_0}$)
- *Refinement stage*:
  $\mathbf{A_{PALM}}$, $\mathbf{S_{PALM}}$ = PALM($\mathbf{X}$,$\mathbf{A_{GMCA}}$,$\mathbf{S_{GMCA}}$)
  The initialization, thresholding strategy and reweighting information come from the warm-up stage.

IOP Conf. Series: Journal of Physics: Conf. Series **1131** (2018) 012007 doi:10.1088/1742-6596/1131/1/012007

Table 1. Average C_A (10 different initializations) for 3 SNR values and 5 algorithms.

	10 dB	15dB	20 dB
2 step	**11.57**	**16.92**	**22.09**
PALM	9.38	10.94	11.01
GMCA	8.87	12.15	15.87
EFICA	−6.92	5.11	9.41
RNA	−6.68	−5.49	6.27

5.2. Empirical results
The experimental protocol is the same as described in sec. 4.2 except that the 2-step algorithm is used instead of PALM. The results are plotted in Fig. 2. With values of C_A higher than 33 dB, the demixing is very accurate. Compared to PALM only, the variance of results over different initializations is also much decreased, becoming almost zero, which shows the robustness of the algorithm with regards to the initialization.

6. Experiment on real data
6.1. Data description and experimental protocol
The goal of this part is to apply our algorithm to realistic data to show its efficiency. The $n = 2$ sources come from simulations obtained from real data of Cassiopeia A supernova remnant. As displayed in Fig. 3, they each consists in a 2D image of resolution $t = 128 \times 128$ pixels, supposed to be approximately sparse in the starlet domain. The mixing matrix is simulated as described in 3.1. To increase the realism of the data and further test the algorithm, we tried three relatively low SNR values: 10, 15 and 20 dB. k is set to 3, which would correspond to a classical hypothesis in terms of Gaussian noise removal.

6.2. Empirical results
The mean of C_A values is displayed in Table 1 and the estimation for a SNR of 15 dB is shown in Fig. 3. The 2-step approach always achieve better results than the two classical BSS algorithms with which we performed the comparison, namely Relative Newton Algorithm (RNA) and Eficient FastICA (EFICA). It also outerpeforms both GMCA and PALM, being always better by at least 2 dB than the best of them.
In addition to the results displayed in Fig. (1), the standard deviation of C_A over different initializations is almost 0, which shows the robustness of the algorithm.

Conclusion
In this work, we show that the ability of recent and theoretically grounded optimization strategies like PALM to provide solutions to sparse BSS problems is highly sensitive to both the initialization and the values of the regularization parameters. To further design an effective as well as robust algorithm, we introduce a 2-step strategy combining PALM with robust heuristic methods such as GMCA. Beyond improving the robustness of PALM-based implementations with respect to initialization, the regularization parameters can be automatically set in the proposed approach. Numerical experiments on both simulated and realistic data demonstrate a high separation quality and good robustness on mixings with low SNR.

Acknowledgment
This work is supported by the European Community through the grant LENA (ERC StG - contract no. 678282).

IOP Conf. Series: Journal of Physics: Conf. Series **1131** (2018) 012007 doi:10.1088/1742-6596/1131/1/012007

References

[1] Comon P and Jutten C 2010 *Handbook of Blind Source Separation: Independent component analysis and applications* (Academic Press)
[2] Gillis N and Glineur F 2012 *Neural Computation* **24** 1085–1105
[3] Bobin J, Starck J L, Moudden Y and Fadili M J 2008 *Advances in Imaging and Electron Physics* **152** 221–302
[4] Zibulevsky M and Pearlmutter B A 2001 *Neural Computation* **13** 863–882
[5] Bronstein A M, Bronstein M M, Zibulevsky M and Zeevi Y Y 2005 *International Journal of Imaging Systems and Technology* **15** 84–91
[6] Li Y, Amari S I, Cichocki A, Ho D W and Xie S 2006 *IEEE Transactions on Signal Processing* **54** 423–437
[7] Candes E J, Wakin M B and Boyd S P 2008 *Journal of Fourier analysis and applications* **14** 877–905
[8] Tseng P 2001 *Journal of Optimization Theory and Applications* **109** 475–494
[9] Bolte J, Sabach S and Teboulle M 2014 *Mathematical Programming* **146** 459–494
[10] Bobin J, Rapin J, Larue A and Starck J L 2015 *IEEE Transanctions on Signal Processing* **63** 1199–1213

Sub-wavelength sensing of bi-periodic materials using topological sensitivity of second-order homogenized model

Marc Bonnet[1], Rémi Cornaggia[2] and Bojan B Guzina[3]

[1] POems (UMR 7231 CNRS-INRIA-ENSTA), ENSTA, Palaiseau, France
[2] Univ Rennes, CNRS, IRMAR - UMR 6625, F-35000 Rennes, France
[3] Department of Civil, Environmental and Geo-Engineering, University of Minnesota, Twin Cities, USA

E-mail: `remi.cornaggia@univ-rennes1.fr`

Abstract. We aim to detect defects or perturbations of periodic media, e.g. due to a defective manufacturing process. To this end, we consider scalar waves in such media through the lens of a second-order macroscopic description, and we compute the sensitivities of the germane effective parameters due to topological perturbations of a microscopic unit cell. Specifically, our analysis focuses on the tensorial coefficients in the governing mean-field equation – including both the leading order (i.e. quasi-static) terms, and their second-order companions bearing the effects of incipient wave dispersion. Then, we present a method that permits sub-wavelength sensing of periodic media, given the (anisotropic) phase velocity of plane waves illuminating the considered medium for several angles and wavenumbers.

1. Introduction

The sensitivity of an homogeneous material to the nucleation of a periodic array of inhomogeneities, and the computation of the corresponding homogenized coefficient, are well-addressed issues, see e.g. [1] and the references therein. By contrast, the sensitivity of an *already periodic* material to an additional periodic perturbation has been studied quite recently, and is mostly used to direct optimization procedures, see e.g. [4] for in-plane elastostatics. Moreover, the focus is often put on static (or quasi-static) configurations, for which the leading-order homogenized coefficients provides a valuable insight of the macroscopic behavior of the fields of interest.

In this study, we assume time-harmonic conditions, and aim at performing *sub-wavelength* sensing of periodic media, i.e. using low-frequency probing waves to detect defects or perturbations of the medium. This goal is achieved by relying on the *anisotropic dispersive properties* of such medium. To capture these properties, we chose to deploy the *second-order model* obtained by two-scale asymptotic homogenization [5], and we compute the sensitivities of the (tensor-valued) coefficients of the corresponding wave equation.

This model, and the corresponding sensitivities, are presented in Section 2. A sub-wavelength sensing method is then developed in Section 3, and applied to the probing of a defective checkerboard-like material. These results are finally summarized in Section 4.

IOP Conf. Series: Journal of Physics: Conf. Series **1131** (2018) 012008 doi:10.1088/1742-6596/1131/1/012008

2. Topological sensitivity of second-order homogenized model

Consider an elastic material occupying a 2D domain and characterized by periodic shear modulus μ and density ρ. The unit cell Y has characteristic length ℓ. Under time-harmonic conditions at circular frequency ω, the antiplane displacement u satisfies the wave equation:

$$\boldsymbol{\nabla} \cdot (\mu \boldsymbol{\nabla} u) + \omega^2 \rho u = 0$$

For long-wavelength configurations (i.e. $\ell \ll \lambda$), two-scale periodic homogenization of this equation [5] leads to the equation satisfied by the mean field U:

$$\boldsymbol{\mu}^0 : \boldsymbol{\nabla}^2 U + \omega^2 \varrho^0 U = -\ell^2 \left[\boldsymbol{\mu}^2 : \boldsymbol{\nabla}^4 U + \omega^2 \boldsymbol{\varrho}^2 : \boldsymbol{\nabla}^2 U \right] + o(\ell^3), \tag{1}$$

where the leading-order and second-order homogenized coefficients $(\boldsymbol{\mu}^0, \varrho^0, \boldsymbol{\mu}^2, \boldsymbol{\varrho}^2)$ are constant tensors, $\boldsymbol{\nabla}^k U$ stands for the k-th gradient of U and ":" denote the inner product of two tensors (i.e. the full contraction over all their indices).

This study considers a periodic perturbation of this material, whereby a penetrable inhomogeneity B_a, of size a and shape $\mathcal{B}$, characterized by contrasts $(\Delta\mu, \Delta\rho)$ is introduced at point $\boldsymbol{z} \in Y$ (Fig. 1). Then, the leading-order expansion of the perturbed coefficients $(\boldsymbol{\mu}_a^0, \varrho_a^0, \boldsymbol{\mu}_a^2, \boldsymbol{\varrho}_a^2)$ w.r.t. a are computed, following previous studies e.g. [4] for in-plane elastostatics.

2.1. *Leading-order coefficients*

Let $\langle \cdot \rangle = |Y|^{-1} \int_Y \cdot$ denote an average on the unit cell. The homogenized density ϱ^0 is defined by $\varrho^0 = \langle \rho \rangle$, so that the perturbed coefficient ϱ_a^0 and the *topological sensitivity* $\mathcal{D}\varrho^0$ are exactly given by:

$$\varrho_a^0 = \varrho^0 + a^2 |Y|^{-1} \mathcal{D}\varrho^0, \quad \text{with } \mathcal{D}\varrho^0 = |\mathcal{B}| \Delta\rho.$$

The homogenized shear modulus $\boldsymbol{\mu}^0$ is defined by $\boldsymbol{\mu}^0 = \langle \mu(\boldsymbol{I} + \boldsymbol{\nabla P}) \rangle^{\mathrm{S}}$, where $\boldsymbol{I}$ is the identity tensor, the first *cell function* $\boldsymbol{P}$ [5] is the Y-periodic and zero-mean vector-valued solution of:

$$\boldsymbol{\nabla} \cdot (\mu(\boldsymbol{I} + \boldsymbol{\nabla P})) = \boldsymbol{0} \tag{2}$$

and the superscript $\cdot^{\mathrm{S}}$ means symmetrization w.r.t. all index permutations. Consequently, $\boldsymbol{\mu}_a^0$ is computed as:

$$\boldsymbol{\mu}_a^0 = \boldsymbol{\mu}^0 + \langle \mu \boldsymbol{\nabla p}_a \rangle^{\mathrm{S}} + \langle \chi_{B_a} \Delta\mu (\boldsymbol{I} + \boldsymbol{\nabla P}_a) \rangle^{\mathrm{S}}$$

where $\boldsymbol{p}_a := \boldsymbol{P}_a - \boldsymbol{P}$ is the perturbation of $\boldsymbol{P}$ and χ_{B_a} is the characteristic function of B_a. The analysis of this perturbation is done by reformulating problem (2) and its perturbed counterpart using domain integral equations, in the spirit of e.g. [2]. With the help of the adjoint state method, it leads to the following leading-order expansion:

$$\boldsymbol{\mu}_a^0 = \boldsymbol{\mu}^0 + a^2 |Y|^{-1} \mathcal{D}\boldsymbol{\mu}^0(\boldsymbol{z}) + o(a^2 |Y|^{-1}), \tag{3}$$

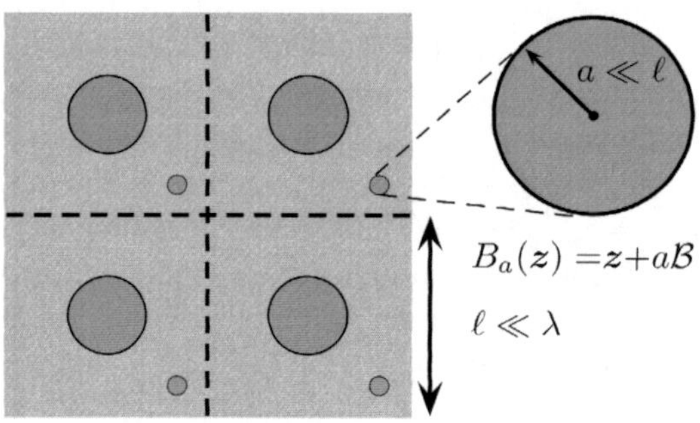

Figure 1: *Notation for a bi-periodic material whose unit cell is perturbed by an inhomogeneity* B_a

IOP Conf. Series: Journal of Physics: Conf. Series **1131** (2018) 012008 doi:10.1088/1742-6596/1131/1/012008

with the topological derivative $\mathcal{D}\boldsymbol{\mu}^0$ given by:

$$\mathcal{D}\boldsymbol{\mu}^0(\boldsymbol{z}) = \big[(\boldsymbol{I} + \boldsymbol{\nabla P}) \cdot \boldsymbol{A} \cdot (\boldsymbol{I} + \boldsymbol{\nabla P})^{\mathrm{T}}\big]\,(\boldsymbol{z}) \tag{4}$$

where $\boldsymbol{A}(\boldsymbol{z}) = \boldsymbol{A}(\mathcal{B}, \mu(\boldsymbol{z}), \Delta\mu)$ is the *polarization tensor* [1] associated to shape $\mathcal{B}$ and moduli $\mu(\boldsymbol{z})$ and $\mu(\boldsymbol{z})+\Delta\mu$. Under notational adjustments, this result is similar to [4]. For homogeneous background materials, in which case $\boldsymbol{P} = \boldsymbol{0}$, one obtains $\mathcal{D}\boldsymbol{\mu}^0 = \boldsymbol{A}$ as shown by [1].

2.2. Second-order coefficients

The second-order homogenized density is defined by $\boldsymbol{\varrho}^2 = \langle \rho \boldsymbol{Q}\rangle^{\mathrm{S}}$, where the *second cell function* $\boldsymbol{Q}$ is the Y-periodic, zero-mean, tensor-valued solution of:

$$\boldsymbol{\nabla}\cdot(\mu(\boldsymbol{P}\otimes\boldsymbol{I}+\boldsymbol{\nabla Q})) = -\mu(\boldsymbol{I}+\boldsymbol{\nabla P}) + (\rho/\varrho^0)\boldsymbol{\mu}^0 \tag{5}$$

Relying on the same integral equation framework, and with careful analysis of the influence of the source terms involving $\boldsymbol{P}_a$ when adressing the perturbed cell function $\boldsymbol{Q}_a$, we show that $\boldsymbol{\varrho}^2_a$ has an expansion of the same form as (3), with its topological derivative $\mathcal{D}\boldsymbol{\varrho}^2$ given by:

$$\mathcal{D}\boldsymbol{\varrho}^2(\boldsymbol{z}) = \Big[(\boldsymbol{I} + \boldsymbol{\nabla P}) \cdot \boldsymbol{A} \cdot \big(\beta\boldsymbol{I} + \boldsymbol{\nabla}\hat{\boldsymbol{X}}[\beta]\big)^{\mathrm{T}} - (\boldsymbol{P}\otimes\boldsymbol{I} + \boldsymbol{\nabla Q}) \cdot \boldsymbol{A} \cdot \boldsymbol{\nabla}\beta$$
$$- \big(\mathcal{D}\boldsymbol{\mu}^0 - (\mathcal{D}\varrho^0/\varrho^0)\boldsymbol{\mu}^0\big)\,\langle \rho(\beta/\varrho^0)\rangle - \mathcal{D}\varrho^0\,\big((\beta/\varrho^0)\boldsymbol{\mu}^0 - \boldsymbol{Q}\big)\Big]^{\mathrm{S}}(\boldsymbol{z}). \tag{6}$$

The above expression features (i) various combinations of the previously computed cell solutions and topological derivatives and (ii) two new *adjoint fields* β and $\hat{\boldsymbol{X}}[\beta]$ defined as the (Y-periodic, zero-mean) solutions of:

$$\boldsymbol{\nabla}\cdot(\mu\boldsymbol{\nabla}\beta) = -(\rho - \varrho^0) \quad \text{and} \quad \boldsymbol{\nabla}\cdot(\mu(\beta\boldsymbol{I}+\boldsymbol{\nabla}\hat{\boldsymbol{X}}[\beta])) = -\mu\boldsymbol{\nabla}\beta. \tag{7}$$

The second-order homogenized shear modulus is defined by $\boldsymbol{\mu}^2 = \langle \mu(\boldsymbol{Q}\otimes\boldsymbol{I} + \boldsymbol{\nabla R})\rangle^{\mathrm{S}}$ in terms of $\boldsymbol{Q}$ and a third cell function $\boldsymbol{R}$, the Y-periodic, zero-mean solution of:

$$\boldsymbol{\nabla}\cdot(\mu(\boldsymbol{Q}\otimes\boldsymbol{I}+\boldsymbol{\nabla R})) = -\mu(\boldsymbol{I}\otimes\boldsymbol{P}+\boldsymbol{\nabla Q}) + (\rho/\varrho^0)\boldsymbol{P}\otimes\boldsymbol{\mu}^0 \tag{8}$$

Once again, an analysis of the problems satisfied by $\boldsymbol{R}$ and $\boldsymbol{R}_a$ is conducted. As a result, $\boldsymbol{\mu}^2_a$ is found to have an expansion similar to (3), and its topological derivative $\mathcal{D}\boldsymbol{\mu}^2$ is:

$$\mathcal{D}\boldsymbol{\mu}^2(\boldsymbol{z}) = \Big\{2\,(\boldsymbol{\nabla P}+\boldsymbol{I})^{\mathrm{T}}\cdot\boldsymbol{A}\cdot(\boldsymbol{\nabla R}+\boldsymbol{I}\otimes\boldsymbol{Q}) - (\boldsymbol{\nabla Q}+\boldsymbol{I}\otimes\boldsymbol{P})^{\mathrm{T}}\cdot\boldsymbol{A}\cdot(\boldsymbol{\nabla Q}+\boldsymbol{I}\otimes\boldsymbol{P})\Big\}^{\mathrm{S}}(\boldsymbol{z})$$
$$+\Big\{\big[\mathcal{D}\boldsymbol{\varrho}^2 + \mathcal{D}\varrho^0\,(\boldsymbol{P}\otimes\boldsymbol{P} - 2\boldsymbol{Q})\big]\otimes\frac{\boldsymbol{\mu}^0}{\varrho^0} + \frac{1}{\varrho^0}\,(\langle\rho\boldsymbol{P}\otimes\boldsymbol{P}\rangle - \boldsymbol{\varrho}^2)\otimes\Big(\mathcal{D}\boldsymbol{\mu}^0 - \mathcal{D}\varrho^0\frac{\boldsymbol{\mu}^0}{\varrho^0}\Big)\Big]^{\mathrm{S}}(\boldsymbol{z}). \tag{9}$$

Remark that the fields $(\boldsymbol{P}, \boldsymbol{Q}, \boldsymbol{R}, \beta, \hat{\boldsymbol{X}}[\beta])$, which are involved in the expressions (4,6,9) of the topological sensitivities, solve the problems (2,5,8,7) which are all posed on the *unperturbed* cell.

3. Sub-wavelength sensing of periodic materials

We now present a *defect identification* procedure, where a "defective" checkerboard material (hereon called the *manufactured* material) is interrogated by plane waves. The sensory data, used to probe for defects, are the phase velocities of such waves for several incidence directions, as captured over a low-frequency (and thus long-wavelength) range. In the identification procedure, the experimental phase velocities are compared to their reference values, computed for the *designed* material. The observed anisotropic dispersion, a key feature of the sensory data, is captured only via *second-order* homogenization, so that both zero-order ($\mathcal{D}\varrho^0$, $\mathcal{D}\boldsymbol{\mu}^0$) and second-order ($\mathcal{D}\boldsymbol{\varrho}^2$, $\mathcal{D}\boldsymbol{\mu}^2$) coefficient sensitivities are needed to interpret the data.

IOP Conf. Series: Journal of Physics: Conf. Series **1131** (2018) 012008 doi:10.1088/1742-6596/1131/1/012008

3.1. Topological sensitivity of the phase velocity

We first recall the second-order, mean field equation

$$\boldsymbol{\mu}^0 : \boldsymbol{\nabla}^2 U + \ell^2 \boldsymbol{\mu}^2 : \boldsymbol{\nabla}^4 U + \omega^2 \left(\varrho^0 U + \ell^2 \boldsymbol{\varrho}^2 : \boldsymbol{\nabla}^2 U \right) = o(\ell^3), \tag{10}$$

according to (1). To characterize the dispersion of a homogenized material, we set the mean field as a plane wave propagating in direction $\boldsymbol{d} = (\cos\theta, \sin\theta)$ with wavenumber k, i.e. $U(\boldsymbol{x}) \propto e^{ik\boldsymbol{d}\cdot\boldsymbol{x}}$. On substituting this plane wave into (10) and neglecting the $o(\ell^3)$ remainder, one obtains the characteristic relation

$$\ell^2 (\boldsymbol{\mu}^2 : \boldsymbol{d}^{\otimes 4}) k^4 - \left((\boldsymbol{\mu}^0 + \ell^2 \omega^2 \boldsymbol{\varrho}^2) : \boldsymbol{d}^{\otimes 2} \right) k^2 + \omega^2 \varrho^0 = 0, \tag{11}$$

(with $\boldsymbol{d}^{\otimes 2} := \boldsymbol{d} \otimes \boldsymbol{d}$ and $\boldsymbol{d}^{\otimes 4} := \boldsymbol{d} \otimes \boldsymbol{d} \otimes \boldsymbol{d} \otimes \boldsymbol{d}$) which in turn yields the anisotropic dispersion formula

$$c(k, \boldsymbol{d}) = \frac{\omega(k, \boldsymbol{d})}{k} = \left(\frac{\boldsymbol{\mu}^0 : \boldsymbol{d}^{\otimes 2} - \ell^2 k^2 \boldsymbol{\mu}^2 : \boldsymbol{d}^{\otimes 4}}{\varrho^0 - \ell^2 k^2 \boldsymbol{\varrho}^2 : \boldsymbol{d}^{\otimes 2}} \right)^{1/2}. \tag{12}$$

For perturbed unit cells, the usual expansion $c_a = c + a^2 |Y|^{-1} \mathcal{D}c + o(a^2)$ holds, and $\mathcal{D}c$ is expressed in terms of the sensitivities of the homogenized coefficients:

$$\mathcal{D}c(k, \boldsymbol{d}, \boldsymbol{z}) = \frac{\left[\mathcal{D}\boldsymbol{\mu}^0 : \boldsymbol{d}^{\otimes 2} - \ell^2 k^2 \mathcal{D}\boldsymbol{\mu}^2 : \boldsymbol{d}^{\otimes 4} - (c(k, \boldsymbol{d}))^2 \left(\mathcal{D}\varrho^0 - \ell^2 k^2 \mathcal{D}\boldsymbol{\varrho}^2 : \boldsymbol{d}^{\otimes 2} \right) \right](\boldsymbol{z})}{2 \left(\varrho^0 - \ell^2 k^2 \boldsymbol{\varrho}^2 : \boldsymbol{d}^{\otimes 2} \right)}. \tag{13}$$

3.2. Quasi-static and dynamic misfit functionals

In what follows, the manufactured material is illuminated by plane waves with wavenumbers $(k_p)_{p=1..N_k}$ propagating in directions $(\theta_j)_{j=1..N_\theta}$. The sensory data are thus the phase velocities $c^{\text{obs}}(k_p, \boldsymbol{d}_j)$, where $\boldsymbol{d}_j = (\cos\theta_j, \sin\theta_j)$. For given wavenumber and direction of incidence $(k, \boldsymbol{d})$, we define the cost functional

$$J(k, \boldsymbol{d}) = \frac{1}{2} \left[c(k, \boldsymbol{d}) - c^{\text{obs}}(k, \boldsymbol{d}) \right]^2, \tag{14}$$

quantifying the misfit between the designed and observed phase velocities. We also define the dynamic cost functional J^{dyn}, extracting the effects of dispersion, as

$$J^{\text{dyn}}(k, \boldsymbol{d}) = \frac{1}{2} \left[\Delta c(k, \boldsymbol{d}) - \Delta c^{\text{obs}}(k, \boldsymbol{d}) \right]^2, \qquad \Delta c(k, \boldsymbol{d}) = c(k, \boldsymbol{d}) - c(k_{\text{min}}, \boldsymbol{d}), \tag{15}$$

where k_{min} is the smallest observed wavenumber (typically, $k_{\text{min}} = k_1 > 0$). The topological sensitivities of these two misfit functionals depend on the sensitivity $\mathcal{D}c$ given by (13) as

$$\begin{aligned} \mathcal{D}J(k, \boldsymbol{d}, \boldsymbol{z}) &= \left(c(k, \boldsymbol{d}) - c^{\text{obs}}(k, \boldsymbol{d}) \right) \mathcal{D}c(k, \boldsymbol{d}, \boldsymbol{z}), \\ \mathcal{D}J^{\text{dyn}}(k, \boldsymbol{d}, \boldsymbol{z}) &= \left(\Delta c(k, \boldsymbol{d}) - \Delta c^{\text{obs}}(k, \boldsymbol{d}) \right) \left(\mathcal{D}c(k, \boldsymbol{d}, \boldsymbol{z}) - \mathcal{D}c(k_{\text{min}}, \boldsymbol{d}, \boldsymbol{z}) \right), \end{aligned} \tag{16}$$

In this setting, we finally define the aggregate *quasistatic* and *dynamic* cost functionals as

$$\mathcal{J}^{\text{stat}} = \sum_{j=1}^{N_\theta} J(k_{\text{min}}, \boldsymbol{d}_j) \qquad \text{and} \qquad \mathcal{J}^{\text{dyn}} = \sum_{j=1}^{N_\theta} \sum_{p=1}^{N_k} J^{\text{dyn}}(k_p, \boldsymbol{d}_j), \tag{17}$$

and their sensitivities $\mathcal{D}\mathcal{J}^{\text{stat}}$ and $\mathcal{D}\mathcal{J}^{\text{dyn}}$ are computed by summing accordingly the sensitivities $\mathcal{D}J$ and $\mathcal{D}J^{\text{dyn}}$ defined by (16).

3.3. Example: incorrectly manufactured chessboard-like material

The chessboard-like material, *as designed*, is depicted in the left panel of Fig. 2, featuring the coefficient ratios $\mu_{\text{max}} = 7\mu_{\text{min}}$ and $\rho_{\text{max}} = 1.2\rho_{\text{min}}$. For this configuration, the coefficients of homogenization are computed using the finite element platform FreeFem++ [3] as:

$$
\begin{aligned}
&\varrho^0 = 1.1\rho_{\text{min}} && \boldsymbol{\mu}^0 = 2.65\mu_{\text{min}}\boldsymbol{I} \\
&\boldsymbol{\varrho}^2 = 1.15\times 10^{-4}\rho_{\text{min}}\boldsymbol{I} && \boldsymbol{\mu}^2 : \begin{cases} \mu^2_{1111} = \mu^2_{2222} = 4.36\times 10^{-3}\mu_{\text{min}} \\ \mu^2_{1122} = 1.20\times 10^{-2}\mu_{\text{min}} \\ \mu^2_{1112} = \mu^2_{1222} = 0 \end{cases}
\end{aligned} \quad . \tag{18}
$$

The chessboard-like material, *as manufactured*, has a defective top-left box in each unit cell, where $\mu = 4$ instead of $\mu = 7$ as shown in the right panel of Fig. 2.

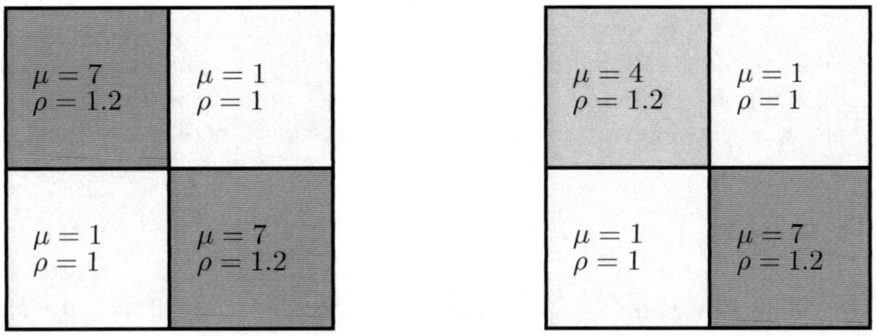

Figure 2: *Designed (left) and manufactured (right)* 1×1 *unit cell of a chessboard-like material.*

To identify and localize the defect, the plane-wave probing grid has $N_\theta = 3$ incident directions $(\theta_1, \theta_2, \theta_3) = (-\pi/4, -\pi/8, 0)$ and $N_k = 10$ wavenumbers $k_p = 2p\,\pi/30$, $p = \overline{1, N_k}$. With such hypotheses, the *shortest wavelength* used to probe the periodic structure is $\lambda_{\text{min}} = 2\pi/k_{N_k} = 3$ which, relative to the 1×1 size of the chosen unit cell, implies sensing *below* the classical diffraction limit. With reference to the above sensing grid, the left panel in Fig. 3 compares the second-order approximation of the phase velocity (12) with the *numerical* values of $c(k, \boldsymbol{d})$ – as computed via Floquet-Bloch transform [5] – for both the designed and manufactured unit cell configuration. As can be seen from the display, the second-order approximation of anisotropic dispersion for the designed material agrees reasonably well with numerical simulations, noting that the small discrepancy between the two is attributed primarily to a limited accuracy of the numerical (Floquet-Bloch) solution due to a combination of material discontinuities inside the unit cell (which slow down the numerical convergence) and computer memory limitations. In contrast, there is a notable discrepancy between the dispersive characteristics of the designed and manufactured material, which justifies the use of the quasistatic functional $\mathcal{J}^{\text{stat}}$ (17) as a basis for (periodic) defect identification. From the results for $\Delta c(k, \boldsymbol{d})$ shown in the right panel of Fig. 3, this is also the case with the dynamic functional $\mathcal{J}^{\text{dyn}}$.

In this setting, the quasistatic sensitivity $\mathcal{DJ}^{\text{stat}}$ serves as an indicator of the relative stiffness of the manufactured material (stiffer or softer than the designed material), via the sign of $\Delta\mu$ which gives $\mathcal{DJ}^{\text{stat}} < 0$. In this vein, Fig. 4 plots $\mathcal{DJ}^{\text{stat}}(\boldsymbol{z})$ over the support of the unit cell, assuming both $\Delta\mu = 1$ (left panel) and $\Delta\mu = -1$ (right panel). To simplify the discussion, the analysis assumes prior knowledge of the fact that the mass density of the unit cell is manufactured exactly (see Fig. 2) by letting $\Delta\rho = 0$. For $\Delta\mu = 1$, we see that $\mathcal{DJ}^{\text{stat}} \geqslant 0$ everywhere; in other words, adding a stiff inclusion to the designed unit cell anywhere would only increase the quasistatic misfit functional. On the contrary, for $\Delta\mu = -1$ one has $\mathcal{DJ}^{\text{stat}} \leqslant 0$ everywhere

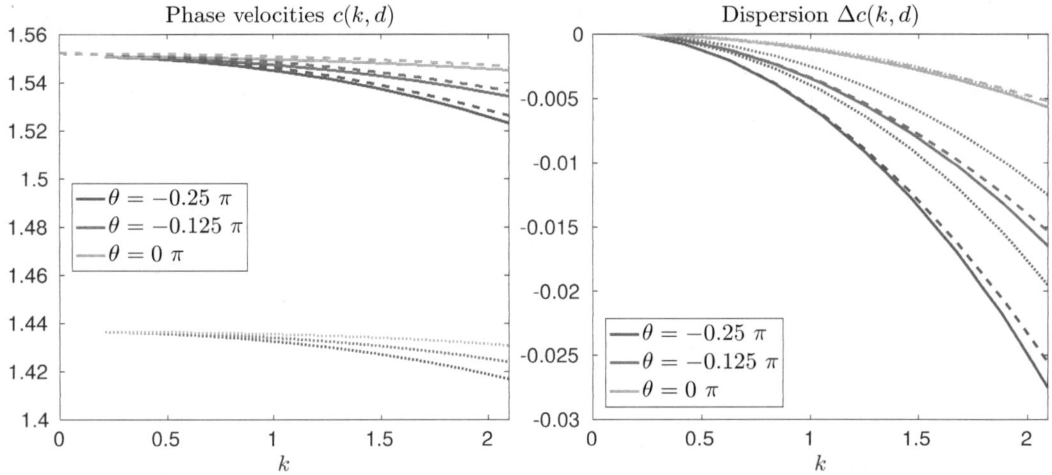

Figure 3: *Left: phase velocity $c(k, \boldsymbol{d})$ for the* designed *chessboard material (homogenized model - dashed, numerical values - solid), and* manufactured *chessboard material (numerical values - dotted). Right: anisotropic dispersion shown in terms of $\Delta c(k, \boldsymbol{d}) = c(k, \boldsymbol{d}) - c(k_{\min}, \boldsymbol{d})$.*

as expected since the manufactured material is *softer* than designed. In light of its sign semi-definiteness for given $\Delta\mu$, however, $\mathcal{DJ}^{\text{stat}}$ appears to have *no localizing capabilities* in that it cannot locate (even approximately) the support of a periodic defect inside the unit cell.

To tackle the latter drawback, we deploy the dynamic sensitivity $\mathcal{DJ}^{\text{dyn}}$ as a sensing lense in Fig. 5, taking $\Delta\mu = -1$ thanks to the quasi-static result. Indeed, the dynamic sensitivity appears to serve as a *remarkable defect locator*: $\mathcal{DJ}^{\text{dyn}} \geqslant 0$ over most of the intact quarter-cells – which are manufactured correctly (except near material interfaces, were the sensitivities tend to localize), while $\mathcal{DJ}^{\text{dyn}} \leqslant 0$ over the entire "defective" upper-left quarter-cell as expected, but also over the lower-right quarter-cell. The latter, however, should not be surprising, since (for the manufactured material at hand) the support of the unit cell could be chosen such that the defective quarter-cell appears as either upper-left or lower-right.

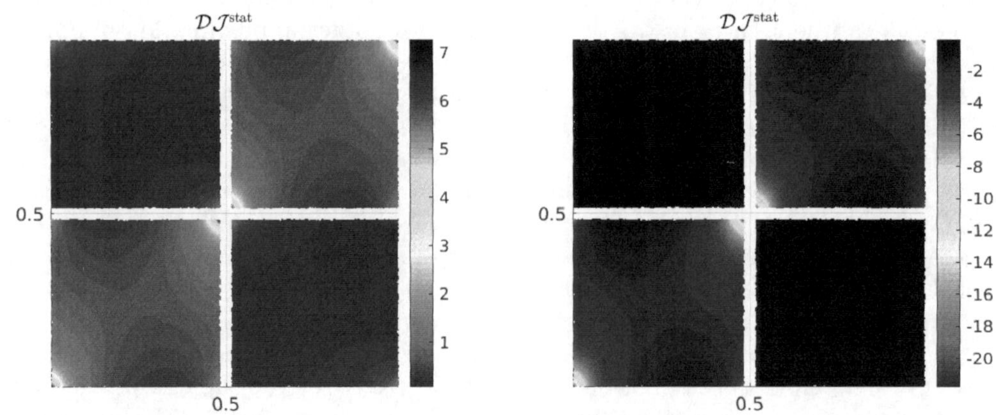

Figure 4: *Distribution of quasistatic sensitivity $\mathcal{DJ}^{\text{stat}}(\boldsymbol{z})$ over the unit cell assuming $\Delta\mu = 1$ (left) and $\Delta\mu = -1$ (right). The near-interface points are omitted from the plot.*

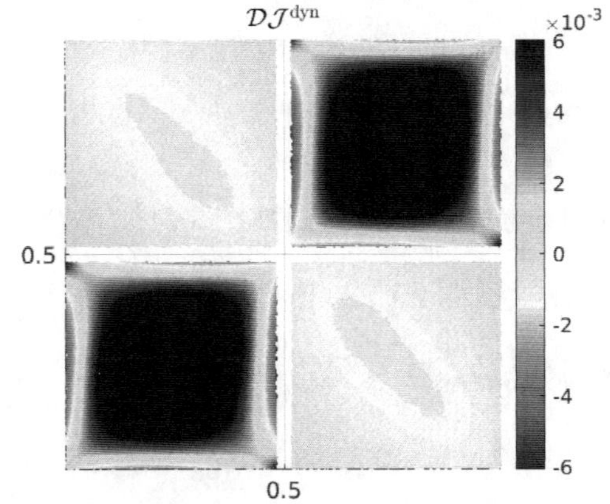

Figure 5: *Distribution of dynamic sensitivity $\mathcal{DJ}^{\text{dyn}}(\boldsymbol{z})$ over the unit cell assuming $\Delta\mu = -1$. The near-interface points are omitted from the plot, and a threshold is applied to the most positive TS values (as only their sign is of interest).*

4. Summary and discussion

We derived the topological sensitivities of the coefficients of the wave equation obtained by second-order homogenization of a periodic material, with respect to the size of an inhomogeneity perturbing the unit cell that defines such material. These sensitivities are expressed in terms of (i) three unit-cell solutions used to formulate the unperturbed macroscopic model, (ii) two adjoint-field solutions driven by the mass density variation inside the unperturbed unit cell, and (iii) the polarization tensor, that synthesizes the geometric and constitutive features of the perturbation.

In light of the above results, the long-wavelength sensing of periodic defects or perturbations could be established by (i) considering the *anisotropic dispersion*-based cost functionals (17), (ii) choosing the sign of the stiffness contrast so that $\mathcal{DJ}^{\text{stat}}(\boldsymbol{z}) \leqslant 0$ everywhere, and (iii) identifying the support of a perturbation inside the unit cell via regions where $\mathcal{DJ}^{\text{dyn}}(\boldsymbol{z}) \leqslant 0$.

Reasonable levels of noise on the data should not alter the efficiency of the method, because (i) the cost functionals aggregate the measurements from several configurations, thus an averaging effect, (ii) the topological sensitivities (16) are linear with respect to the data and therefore to the noise and (iii) only the sign (and not the values) of these sensitivities are of interest.

Finally, the choice of illumination by plane waves enabled to retain the phase velocity as a dispersion indicator. If such configuration cannot be reached in practice (e.g. because captors are too close from a source transducer for the plane wave approximation to hold), one should use another indicator to measure the dispersive properties of the material and define new cost functionals accordingly.

References

[1] Ammari H, Kang H and Touibi K 2005 Boundary layer techniques for deriving the effective properties of composite materials. *Asymptot. Anal.* **41** 119–140
[2] Bonnet M 2009 Higher-order topological sensitivity for 2-d potential problems. Application to fast identification of inclusions. *Int. J. Solids Struct.* **46** 2275–2292
[3] Hecht F 2012 New development in FreeFem++. *J. Numer. Math.* **20** 251–265
[4] Giusti SM, Novotny AA and de Souza Neto EA 2010 Sensitivity of the macroscopic response of elastic microstructures to the insertion of inclusions. *Proc. Roy. Soc. London. Series A.*
[5] Wautier A and Guzina BB 2015 On the second-order homogenization of wave motion in periodic media and the sound of a chessboard. *J. Mech. Phys. Solids* **78** 382–404

IOP Conf. Series: Journal of Physics: Conf. Series **1131** (2018) 012009 doi:10.1088/1742-6596/1131/1/012009

Optimality of 1-norm regularization among weighted 1-norms for sparse recovery: a case study on how to find optimal regularizations

Yann Traonmilin[a] and Samuel Vaiter[b]

[a]CNRS, Institut de Mathématiques de Bordeaux, 33400 Talence, France; [b]CNRS, Institut de Mathématiques de Bourgogne, 9 avenue Alain Savary, 21000, Dijon

E-mail: [a]yann.traonmilin@math.u-bordeaux.fr, [b]samuel.vaiter@u-bourgogne.fr

Abstract. The 1-norm was proven to be a good convex regularizer for the recovery of sparse vectors from under-determined linear measurements. It has been shown that with an appropriate measurement operator, a number of measurements of the order of the sparsity of the signal (up to log factors) is sufficient for stable and robust recovery. More recently, it has been shown that such recovery results can be generalized to more general low-dimensional model sets and (convex) regularizers. These results lead to the following question: to recover a given low-dimensional model set from linear measurements, what is the "best" convex regularizer? To approach this problem, we propose a general framework to define several notions of "best regularizer" with respect to a low-dimensional model. We show in the minimal case of sparse recovery in dimension 3 that the 1-norm is optimal for these notions. However, generalization of such results to the n-dimensional case seems out of reach. To tackle this problem, we propose looser notions of best regularizer and show that the 1-norm is optimal among weighted 1-norms for sparse recovery within this framework.

1. Introduction

We consider the observation model in a Hilbert space $\mathcal{H}$ (with associated norm $\|\cdot\|_{\mathcal{H}}$):

$$y = Mx_0 \tag{1}$$

where M is an under-determined linear operator, y is a m-dimensional vector and x_0 is the unknown. We suppose that x_0 belongs to a low-dimensional model Σ (a union of subspaces). We consider the following minimization program.

$$x^* \in \arg \min_{Mx=y} R(x) \tag{2}$$

where R is a regularization function. A huge body of work gives practical regularizers ensuring that $x^* = x_0$ for several low-dimensional models (in particular sparse and low rank models, see [5] for a most complete review of these results). The operator M is generally required to satisfy some property (e.g., the restricted isometry property) to guarantee recovery. In this work, we aim at finding the "best" regularizer for exact recovery of $x_0 \in \Sigma$. Ideally we would like to set $R = \iota_\Sigma$ (the characteristic function of Σ) but it is not practical in many cases (sparse and low rank recovery) as it is generally not convex, and even NP-hard to compute as a combinatorial problem. Consequently, we restrict the search for the best regularizer to a class of interesting regularizers $\mathcal{C}$. In our examples, the set $\mathcal{C}$ is a subset of the set of convex functions. Other interesting classes might be considered, such as partly smooth functions [9].

1.1. Best regularizer with respect to a low dimensional model

Defining what is the "best" regularizer in $\mathcal{C}$ for recovery is not immediate. Ideally, to fit to the inverse problem, we must define a compliance measure that depends on both the kind of unknown and measurement operator we consider. If we have some knowledge that $M \in \mathcal{M}$ where $\mathcal{M}$ is a set of linear operators, we want to define a compliance measure $A_{\Sigma,\mathcal{M}}(R)$ that tells us if a regularizer is good in these situations, and maximize it. Such maximization might yield a function R^* that depends on M (e.g., in [7], when looking for tight continuous relaxation of the ℓ^0 penalty a dependency on M appears). We aim for a more *universal* notion of optimal regularizer that does not depend on M. Hence, we look for a compliance measure $A_\Sigma(R)$ and its maximization

$$R^* \in \arg\max_{R \in \mathcal{C}} A_\Sigma(R). \tag{3}$$

In the sparse recovery example studied in this article, the existence of a maximum of $A_\Sigma(R)$ is verified. However, we could ask ourselves what conditions on $A_\Sigma(R)$ and $\mathcal{C}$ are necessary and sufficient for the existence of a maximum, which is out of the scope of this article.

1.2. Compliance measures

When studying recovery with a regularization function R, two types of guarantees are generally used: uniform and non-uniform. To describe these recovery guarantees, we use the following definition of descent vectors.

Definition 1.1 (Descent vectors). *For any $x \in \mathcal{H}$, the collection of descent vectors of R at x is*

$$\mathcal{T}_R(x) := \{z \in \mathcal{H} : R(x+z) \le R(x)\}. \tag{4}$$

We write $\mathcal{T}_R(\Sigma) := \bigcup_{x\in\Sigma} \mathcal{T}_R(x)$. Recovery is characterized by descent vectors (recall that x^* is the result of minimization (2)):

- Uniform recovery: Let M a linear operator. Then "for all $x_0 \in \Sigma$, $x^* = x_0$" is equivalent to $\mathcal{T}_R(\Sigma) \cap \ker M = \{0\}$.
- Non-uniform recovery: Let M a linear operator and $x_0 \in \Sigma$. Then $x^* = x_0$ is equivalent to $\mathcal{T}_R(x_0) \cap \ker M = \{0\}$.

Hence, a regularization function R is "good" if $\mathcal{T}_R(\Sigma)$ leaves a lot of space for $\ker M$ to not intersect it (trivially). In dimension n, if there is no orientation prior on the kernel of M, the amount of space left can be quantified by the "volume" of $\mathcal{T}_R(\Sigma) \cap S(1)$ where $S(1)$ is the unit sphere with respect to $\|\cdot\|_{\mathcal{H}}$. Hence a compliance measure for uniform recovery can be defined as

$$A_\Sigma^U(R) := 1 - \frac{\mathrm{vol}\,(\mathcal{T}_R(\Sigma) \cap S(1))}{\mathrm{vol}(S(1))}. \tag{5}$$

More precisely, here, the volume $\mathrm{vol}(E)$ of a set E is the measure of E with respect to the uniform measure on the sphere $S(1)$ (i.e. the $n-1$-dimensional Haussdorf measure of $\mathcal{T}_R(\Sigma) \cap S(1)$). When maximizing this compliance measure with convex regularizers (proper, coercive and continuous), it has been shown that we can limit ourselves to atomic norms with atoms included in the model [8, Lemma 2.1]. When looking at non-uniform recovery for random Gaussian measurements, the quantity $\frac{\mathrm{vol}(\mathcal{T}_R(x_0)\cap S(1))}{\mathrm{vol}(S(1))}$ represents the probability that a randomly oriented kernel of dimension 1 intersects (non trivially) $\mathcal{T}_R(x_0)$. The highest probability of intersection with respect to x_0 quantifies the lack of compliance of R, hence we can define:

$$A_\Sigma^{NU}(R) := 1 - \sup_{x\in\Sigma} \frac{\mathrm{vol}\,(\mathcal{T}_R(x) \cap S(1))}{\mathrm{vol}(S(1))} \tag{6}$$

Note that this can be linked with the Gaussian width and statistical dimension theory of sparse recovery [3, 1].

Remark 1.1. *In infinite dimension, the volume of the sphere $S(1)$ vanishes, making the measures above uninteresting. However, [8] and [6] show that we can often come back to a low-dimensional recovery problem in an intermediate finite (potentially high dimensional) subspace of $\mathcal{H}$. Adapting the definition of $S(1)$ to this subspace allows to extend these compliance measures.*

Another possibility for the n-dimensional case, which we develop in this article, is to use recovery results based on the restricted isometry property (RIP). They have been shown to be adequate for multiple models [8], to be tight in some sense [4] for sparse and low rank recovery, to be necessary in some sense [2] and to be well adapted to the study of random operators [6]. In particular, it has been shown that if M has a RIP with constant $\delta < \delta_\Sigma(R)$ on the secant set $\Sigma - \Sigma$, with $\delta_\Sigma(R)$ being fully determined by Σ and R [8], then stable recovery is possible. Hence, by taking

$$A_\Sigma^{RIP}(R) := \delta_\Sigma(R), \tag{7}$$

the larger A_Σ^{RIP} is, the less stringent are RIP recovery conditions for recovery of elements of Σ with R. We develop these ideas more precisely in Section 3 and perform the maximization with such measure in the sparse recovery case.

1.3. Optimality results

Within this framework we show the following results:

(i) In Section 2, when $\mathcal{H} = \mathbb{R}^3$, $\Sigma = \Sigma_1$ the set of 1-sparse vectors and $\mathcal{C}$ is the set of (symmetric) atomic norms with atoms included in the model. Both $A_\Sigma^U(R)$ and $A_\Sigma^{NU}(R)$ are uniquely maximized by multiples of the ℓ^1-norm (Theorem 2.1). In this case, it is possible to exactly compute and maximize the compliance measure. While this study gives a good geometrical insight of the quantities at hand, extending these exact calculations to the general k-sparse recovery in dimension n seems out of reach.

(ii) In Section 3, we describe precisely how compliance measures based on RIP recovery can be defined. We then study two of these measures (based on state of the art recovery guarantees) for $\mathcal{H} = \mathbb{R}^n$, $\Sigma = \Sigma_k$ the set of k-sparse vectors (vectors with at most k non zero coordinates) and $\mathcal{C}$ the set of weighted ℓ^1-norms. We show that both of these measures are uniquely maximized by multiples of the ℓ^1-norm (Theorem 3.1 and Theorem 3.2).

2. The case of 1-sparse vectors in 3D

We investigate in detail the case of 3-dimensional vectors which are 1 sparse, i.e. the simplest interesting case (uniform recovery of 1-sparse vectors in 2 dimensions is impossible if M is not invertible). Here, we have $\Sigma = \Sigma_1$ and $\mathcal{H} = \mathbb{R}^3$. We consider weighted ℓ^1-norms, i.e., atomic norms of the form $\mathcal{C} = \{\|\cdot\|_w : \|z\|_w = w_1|z_1| + w_2|z_2| + w_3|z_3|\} = \{\|\cdot\|_\mathcal{A} : \mathcal{A} = \{\pm w_i^{-1} e_i\}_{i=1,3}, w_i > 0, \max|w_i| = 1\}$ (where $\|\cdot\|_\mathcal{A}$ is the atomic norm generated by atoms $\mathcal{A}$). To maximize A_Σ^U (respectively A_Σ^{NU}), we have to compute the surface of the intersections of 3 descent cones with the unit sphere $S(1)$ (respectively the surface of the biggest possible intersection of a descent cone with $S(1)$).

As shown in Figure 1, with symmetries, we need to compute the intersection of 3 descent cones of a weighted ℓ^1-norm at 1-sparse vectors. This is the object of Lemma 2.1. Note that in the case of asymmetrical atoms, we would then need to calculate 6×4 intersections of the sphere with a tetrahedron. To ease the notations, we introduce the following quantities which represent the cosine of the angles of a tetrahedron of interest. For $i \neq j$, set $\mu_i = w_i^{-1}$,$\beta_{ij}(\mu) = \cos(\alpha_{ij}) = \left(1 + (\mu_j/\mu_i)^2\right)^{-\frac{1}{2}}$ and $\beta_{ii} = \prod_{i \neq j} \beta_{ij}$.

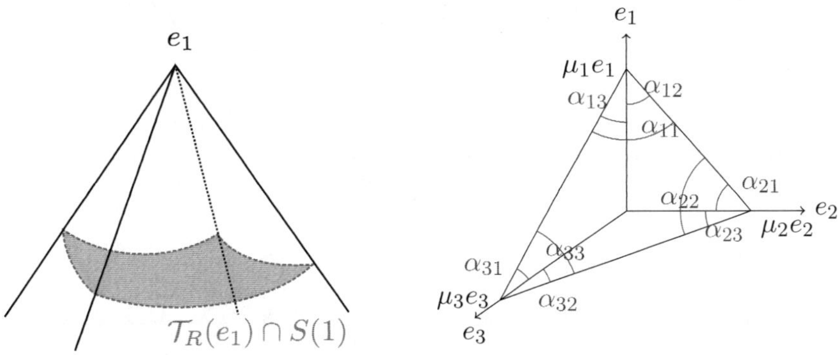

Figure 1. Left: a representation of the surface of interest. Right: parameterization of the angles of a tetrahedron.

Lemma 2.1. *Let $i \in \{1,2,3\}$, $x \in \mathbb{R}e_i$ a 1-sparse vector and $\mu_1, \mu_2, \mu_3 > 0$ three positive real numbers. Then,*

$$\mathrm{vol}\left(\mathcal{T}_{\|\cdot\|_w}(x) \cap S(1)\right) = 4\tan^{-1}\left(\frac{1}{1+c_i(\mu)}\right) \tag{8}$$

where

$$c_i(\mu) = 1 + \sum_{j\neq i} \beta_{ij}(\mu) + \prod_{j\neq i} \beta_{ij}(\mu). \tag{9}$$

With this Lemma 2.1, we can prove that the ℓ^1-norm is the best regularizer for 1-sparse signals both for uniform and non-uniform recovery among all weighted ℓ^1-norms.

Theorem 2.1. *Let $\Sigma = \Sigma_1$ the set of 1-sparse vectors in $\mathbb{R}^3$ and $\mathcal{C} = \{\|\cdot\|_w : \|z\|_w = w_1|z_1| + w_2|z_2| + w_3|z_3|, \max|w_i| = 1, w_i > 0\}$. The ℓ^1-norm is the best regularizer among the class $\mathcal{C}$ for Σ, both for the uniform and non-uniform case,*

$$\arg\max_{R\in\mathcal{C}} A_\Sigma^U(R) = \arg\max_{R\in\mathcal{C}} A_\Sigma^{NU}(R) = \|\cdot\|_1$$

We expect a similar result if $\mathcal{C}$ is generalized to asymmetrical atomic norms. To extend these exact calculations to the n-dimensional case, we would need to be able to compare intersections of spheres and descent cones of R without an analytic formula, which appears to be a very difficult task.

3. Compliance measures based on the RIP

In this Section, we detail how we can use RIP recovery conditions to build compliance measures that are more easily managed. We start by recalling definitions and results about RIP recovery guarantees then apply our methodology. We also give Lemma that emphasize the relevant quantity (depending on the geometry of the regularizer and the model) to optimize.

Definition 3.1 (RIP constant). *In a Hilbert space $\mathcal{H}$, let Σ a union of subspaces. Let M a linear map, the RIP constant of M is defined as*

$$\delta(M) = \inf_{x\in\Sigma-\Sigma}\left|\frac{\|Mx\|_{\mathcal{H}}^2}{\|x\|_{\mathcal{H}}^2} - 1\right|, \tag{10}$$

where $\Sigma - \Sigma$ (differences of elements of Σ) is called the secant set.

In [8], an explicit constant $\delta_\Sigma^{suff}(R)$ is given, such that $\delta(M) < \delta_\Sigma^{suff}(R)$ guarantees exact recovery of elements of Σ by minimization (2). This constant is only sufficient (and sharp in some sense for sparse and low rank recovery). An ideal RIP based compliance measure would be to use a sharp RIP constant $\delta_\Sigma^{sharp}(R)$ (which is not explicit, it is an open question to derive closed form formulations of this constant for sparsity and other low-dimensional models) defined as:

$$\delta_\Sigma^{sharp}(R) := \inf_{M:\ker M \cap \mathcal{T}_R(\Sigma) \neq \{0\}} \delta(M). \tag{11}$$

It is the best RIP constant of measurement operators where uniform recovery fail. The lack of analytic expressions for $\delta_\Sigma^{sharp}(R)$ limits the possibilities of exact optimization with respect to R. We propose to look at two compliance measures:

- Measures based on necessary RIP conditions [4] which yields sharp recovery constants for particular set of operators, e.g.,

$$\delta_\Sigma^{nec}(R) := \inf_{z \in \mathcal{T}_R(\Sigma)\setminus\{0\}} \delta(I - \Pi_z). \tag{12}$$

  where Π_z is the orthogonal projection onto the one-dimensional subspace span(z) (other intermediate necessary RIP constants can be defined). Another open question is to determine whether $\delta_\Sigma^{nec}(R) = \delta_\Sigma^{sharp}(R)$ generally or for some particular models.
- Measures based on sufficient RIP constants for recovery (e.g., $\delta_\Sigma^{suff}(R)$ from [8]).

Note that we have the relation

$$\delta_\Sigma^{suff}(R) \le \delta_\Sigma^{sharp}(R) \le \delta_\Sigma^{nec}(R). \tag{13}$$

To summarize, in the following, instead of considering the most natural RIP based compliance measure (based on $\delta_\Sigma^{sharp}(R)$), we use the best known bounds of this measure.

3.1. Compliance measures based on necessary RIP conditions

In this case, instead of working with actual RIP constants, it is easier to use (equivalently) the restricted conditioning.

Definition 3.2 (Restricted conditioning).

$$\gamma(M) := \frac{\sup_{x \in (\Sigma-\Sigma)\cap S(1)} \|Mx\|_\mathcal{H}^2}{\inf_{x \in (\Sigma-\Sigma)\cap S(1)} \|Mx\|_\mathcal{H}^2}. \tag{14}$$

The RIP constant $\delta(M)$ is increasing with respect to $\gamma(M)$. In the following, we consider the compliance measure

$$A_\Sigma^{RIP,nec}(R) := \gamma_\Sigma(R) = \inf_{z \in \mathcal{T}_R(\Sigma)\setminus\{0\}} \gamma(I - \Pi_z). \tag{15}$$

When maximizing $A_\Sigma^{RIP,nec}(R)$, we look at optimal regularizers for recovery with tight frames with a kernel of dimension 1.

From here, we specialize to $\Sigma = \Sigma_k$ and $\mathcal{H} = \mathbb{R}^n$. Hence $\Sigma - \Sigma = \Sigma_{2k}$ with $k \ge 1$ and $n \ge 3$ (for $n < 3$ uniform recovery is not possible for non-invertible M). To show optimality of the ℓ^1-norm, we use the following characterization of $A_\Sigma^{RIP,nec}(R)$.

Lemma 3.1. *Let $\Sigma = \Sigma_k$ and $\mathcal{H} = \mathbb{R}^n$. Then*

$$A_\Sigma^{RIP,nec}(R) = \frac{1}{1 - \inf_{z \in \mathcal{T}_R(\Sigma)\setminus\{0\}} \sup_{x \in (\Sigma-\Sigma)\cap S(1)} \frac{\langle x,z\rangle^2}{\|z\|_\mathcal{H}^2}}. \tag{16}$$

IOP Conf. Series: Journal of Physics: Conf. Series **1131** (2018) 012009 doi:10.1088/1742-6596/1131/1/012009

We consider the set $\mathcal{C} = \{R : R(x) = \sum w_i|x_i| = \|x\|_w, \max|w_i| = 1, w \in (\mathbb{R}_+^*)^n\}$. Note that

$$\arg\max_{R\in\mathcal{C}} A_\Sigma^{RIP,nec}(R) = \arg\min_{R\in\mathcal{C}} \sup_{z\in\mathcal{T}_R(\Sigma)\setminus\{0\}} \frac{\|z_{T_2^c}\|_2^2}{\|z_{T_2}\|_2^2} = \arg\min_{R\in\mathcal{C}} B_\Sigma(R) \tag{17}$$

where T_2 is a notation for the support of $2k$ biggest coordinates in z, i.e. for all $i \in T_2, j \in T_2^c$, we have $|z_i| \geq |z_j|$.

Studying the quantity $B_\Sigma(R) := \sup_{z\in\mathcal{T}_R(\Sigma)\setminus\{0\}} \frac{\|z_{T_2^c}\|_2^2}{\|z_{T_2}\|_2^2}$ for $R \in \mathcal{C}$ permits to conclude.

Theorem 3.1. *Let $\Sigma = \Sigma_k$, $\mathcal{H} = \mathbb{R}^n$ and $\mathcal{C} = \{R : R(x) = \sum w_i|x_i| = \|x\|_w, \max|w_i| = 1, w \in (\mathbb{R}_+^*)^n\}$. Suppose $n \geq 2k$, then*

$$\|\cdot\|_1 = \arg\max_{R\in\mathcal{C}} A_\Sigma^{RIP,nec}(R). \tag{18}$$

In the next section, we will see that the optimization of the sufficient RIP constant leads to very similar expressions.

3.2. Compliance measures based on sufficient RIP conditions

In [8], it was shown for Σ a union of subspaces and an arbitrary regularizer R, that an explicit RIP constant $\delta_\Sigma^{suff}(R)$ is sufficient to guarantee reconstruction. From here we set $\Sigma := \Sigma_k$, $\mathcal{C} = \{R : R(x) = \sum w_i|x_i| = \|x\|_w, \max|w_i| = 1, w \in (\mathbb{R}_+^*)^n\}$.

Proposition 3.1. *Let $\Sigma = \Sigma_k$ the set of k-sparse vectors. Consider the constant $\delta_\Sigma^{suff}(R)$ from [8][Eq. (5)], we have:*

$$\delta_\Sigma^{suff}(R) = \frac{1}{\sqrt{\sup_{z\in\mathcal{T}_R(\Sigma)\setminus\{0\}} \frac{\|z_{T^c}\|_\Sigma^2}{\|z_T\|_2^2} + 1}} \tag{19}$$

where T denotes the support of the k biggest coordinates of z and $\|\cdot\|_\Sigma$ is the atomic norm generated by $\Sigma \cap S(1)$ (the convex gauge induced by the convex envelope of $\Sigma \cap S(1)$).

In the following, we use

$$A_\Sigma^{RIP,suff}(R) := \delta_\Sigma^{suff}(R). \tag{20}$$

It must be noted that characterization of Proposition 3.1 was used as a lower bound on δ_Σ^{suff} in [8] (hence this part of the proof of the control of δ_Σ^{suff} in [8] is exact).

Similarly to the necessary case, from Proposition 3.1, we have

$$\arg\max_{R\in\mathcal{C}} A_\Sigma^{RIP,suff}(R) = \arg\min_{R\in\mathcal{C}} \sup_{z\in\mathcal{T}_R(\Sigma)\setminus\{0\}} \frac{\|z_{T^c}\|_\Sigma^2}{\|z_T\|_2^2} = \arg\min_{R\in\mathcal{C}} D_\Sigma(R) \tag{21}$$

where T denotes the support of the k biggest coordinates of z and $D_\Sigma(R) := \sup_{z\in\mathcal{T}_R(\Sigma)\setminus\{0\}} \frac{\|z_{T^c}\|_\Sigma^2}{\|z_T\|_2^2}$.

Remark the similarity between the fundamental quantity to optimize for the necessary case and the sufficient case $B_\Sigma(R)$ and $D_\Sigma(R)$. Studying the quantity $D_\Sigma(R)$ for $R \in \mathcal{C}$ leads to the result.

Theorem 3.2. *Let $\Sigma = \Sigma_k$, $\mathcal{H} = \mathbb{R}^n$ and $\mathcal{C} = \{R : R(x) = \sum w_i|x_i| = \|x\|_w, \max|w_i| = 1, w \in (\mathbb{R}_+^*)^n\}$. Suppose $n \geq 2k$, then*

$$\|\cdot\|_1 = \arg\sup_{R\in\mathcal{C}} A_\Sigma^{RIP,suff}(R). \tag{22}$$

4. Discussion and future work

We have shown that, not surprisingly, the ℓ^1-norm is optimal among weighted ℓ^1-norms for sparse recovery for several notions of compliance. This result had to be expected due to symmetries of the problem. However, the important point is that we could explicitly quantify the notion of good regularizer. This is promising for the search of optimal regularizers for more complicated low-dimensional models such as "sparse and low rank" models or hierarchical sparse models. For the case of sparsity, we expect to be able to generalize the optimality of the ℓ^1-norm to the more general case of atomic norms with atoms included in the model. We also expect similar result for low-rank recovery and the nuclear norm as technical tools are very similar.

It must be noted that for RIP compliance measures, we did not use a constructive proof (we exhibited the maximum of the compliance measure). A constructive proof, i.e. an exact calculation and optimization of the quantities $B_\Sigma(R)$ and $D_\Sigma(R)$ would be more satisfying as it would not require the knowledge of the optimum, which is our ultimate objective.

We used compliance measures based on (uniform) RIP recovery guarantees to give results for the general sparse recovery case, it would be interesting to do such analysis using (non-uniform) recovery guarantees based on the statistical dimension or Gaussian width of the descent cones [3, 1]. One would need to precisely lower and upper bound these quantities, similarly to our approach with the RIP, to get satisfying results.

Finally, while these compliance measures are designed to make sense with respect to known results in the area of sparse recovery, one might design other compliance measures tailored for particular needs, in this search for optimal regularizers.

Acknowledgements

This work was partly supported by the CNRS PEPS JC 2018 (project on efficient regularizations). The authors would like to thank Rémi Gribonval for his insights on this problem.

References

[1] D. Amelunxen, M. Lotz, M. B. McCoy, and J. A. Tropp. Living on the edge: phase transitions in convex programs with random data. *Information and Inference*, 3(3):224–294, 2014.

[2] A. Bourrier, M. Davies, T. Peleg, P. Perez, and R. Gribonval. Fundamental performance limits for ideal decoders in high-dimensional linear inverse problems. *Information Theory, IEEE Transactions on*, 60(12):7928–7946, 2014.

[3] V. Chandrasekaran, B. Recht, P. Parrilo, and A. Willsky. The convex geometry of linear inverse problems. *Foundations of Computational Mathematics*, 12(6):805–849, 2012.

[4] M. E. Davies and R. Gribonval. Restricted isometry constants where ℓ^p sparse recovery can fail for $0 < p \leq 1$. *Information Theory, IEEE Transactions on*, 55(5):2203–2214, 2009.

[5] S. Foucart and H. Rauhut. *A mathematical introduction to compressive sensing*. Springer, 2013.

[6] G. Puy, M. E. Davies, and R. Gribonval. Recipes for stable linear embeddings from hilbert spaces to $\mathbb{R}^m$. *arXiv preprint arXiv:1509.06947*, 2015.

[7] E. Soubies, L. Blanc-Féraud, and G. Aubert. A continuous exact ℓ_0 penalty (cel0) for least squares regularized problem. *SIAM Journal on Imaging Sciences*, 8(3):1607–1639, 2015.

[8] Y. Traonmilin and R. Gribonval. Stable recovery of low-dimensional cones in Hilbert spaces: One RIP to rule them all. *Applied And Computational Harmonic Analysis, In Press*, 2016.

[9] S. Vaiter, M. Golbabaee, J. Fadili, and G. Peyré. Model selection with low complexity priors. *Information and Inference: A Journal of the IMA*, 4(3):230–287, 2015.

IOP Conf. Series: Journal of Physics: Conf. Series **1131** (2018) 012010 doi:10.1088/1742-6596/1131/1/012010

Imaging defects in an elastic waveguide using time-dependent surface data

V Baronian[1], L Bourgeois[2], B Chapuis[1] and A Recoquillay[1]

[1] CEA, LIST, Gif-sur-Yvette, France

[2] Laboratoire POEMS, ENSTA ParisTech, Palaiseau, France

E-mail: `arnaud.recoquillay@cea.fr`

Abstract. We are interested here in applying the Linear Sampling Method to the context of Non Destructive Testing for waveguides. Specifically the Linear Sampling Method [1] in its modal form [2] is adapted to image defects in an elastic waveguide from realistic scattering data, that is data coming from sources and receivers on the surface of the waveguide in the time domain, as it has already been done in the acoustic case [16]. The obtained method is applied to artificial data and to experimental data in the special case of back-scattering.

Introduction

This paper presents a Non Destructive Testing method to locate and reconstruct the shape of some defects in an elastic waveguide using guided modes. Since the sensor is supposed to be far from the defect, the data consist in the scattered waves associated with the propagating modes only. The usual methods rely on a small number of propagating modes (at low frequency) [3, 4, 5, 6], which allow the tracking of each of these modes. In addition, these methods depend on the type of defect which is expected. Our method is based on a mathematical approach, namely the Linear Sampling Method (LSM) [1] and more precisely in a modal formulation [2], which takes into account the scattered fields associated with a large number of propagating modes (at high frequency). The LSM consists, for each sampling point z describing a grid, to check if some appropriate and analytically known test function depending on z belongs or not to the range of a linear operator, the definition of which is based on the data. In the case of a positive answer, the point z belongs to the defect, which is a practical way of retrieving its boundary. An important feature of the LSM is that it is independent of the number of components and of the type of the defect.

This method has been widely studied since its introduction [7, 8, 9, 13] but its application has been restricted to artificial data except for electromagnetism, see [12], and most of the time in the frequency regime (for time domain data, see [10, 11]). We propose here to adapt the LSM to realistic data in the non destructive testing framework, that is data in the time domain measured on the surface of the inspected waveguide.

Our strategy is the following. Firstly, we transform the time domain data into multi-frequency data with the help of the Fourier transform. Secondly, at each frequency we transform the surface data into a novel set of ideal data which are suitable for the modal formulation of the LSM. Thirdly, we combine all the images given by the LSM at each frequency in order to obtain the best possible defect identification. In addition, the optimization of the sources and receivers

IOP Conf. Series: Journal of Physics: Conf. Series **1131** (2018) 012010 doi:10.1088/1742-6596/1131/1/012010

is possible thanks to this strategy. The Linear Sampling Method in the frequency domain for an elastic waveguide is already justified for a Dirichlet obstacle in [2] and for cracks in [14]. The justification for a Neumann obstacle would follow the same lines and the justification for a penetrable obstacle would be very close to that of [15] for the periodic acoustic waveguide. Hence, this article is focused on the application of the LSM in the presence of real data (in particular, surface data in the time domain) and on how the substantial issues generated by those real data can be fixed.
The outline of this paper is what follows: the LSM in a modal formulation is briefly summarized in section 1 while the case of data on the surface in the frequency domain is explained in section 2. In section 3 the adaptation of the method to time domain data is shown and an application of the method to artificial data is done. Finally the experimental setting is explained in section 4 and the corresponding imaging results are presented.

1. The Linear Sampling Method: a modal formulation

Let the space dimension be 2. We consider a waveguide $W = \Sigma \times \mathbb{R}$ of transverse section Σ and boundary Γ. This waveguide is made of an isotropic material of density ρ and Lamé constants (λ, μ). Let be denoted $\boldsymbol{u}$ the displacement, σ the stress tensor associated with $\boldsymbol{u}$ following

$$\sigma(\boldsymbol{u}) = \lambda(\mathrm{div}\boldsymbol{u})\, I + \mu(\nabla\boldsymbol{u} + \nabla\boldsymbol{u}^T),$$

and let us decompose

$$\boldsymbol{u} = \begin{pmatrix} \boldsymbol{u}_S \\ u_3 \end{pmatrix}, \qquad \sigma \cdot \boldsymbol{e}_3 = \begin{pmatrix} \boldsymbol{t}_S \\ -t_3 \end{pmatrix},$$

where the subscripts S and 3 denote the components of a vector along the transverse section and along the axis, respectively. We introduce the mixed variables $\boldsymbol{X}$, $\boldsymbol{Y}$ defined by:

$$\boldsymbol{X} = \begin{pmatrix} \boldsymbol{t}_S \\ u_3 \end{pmatrix}, \qquad \boldsymbol{Y} = \begin{pmatrix} \boldsymbol{u}_S \\ t_3 \end{pmatrix}.$$

The guided modes are the solutions with separated variables of

$$\begin{cases} \mathrm{div}\sigma(\boldsymbol{u}) + \rho\omega^2\boldsymbol{u} = 0 & \text{in } W \\ \sigma(\boldsymbol{u})\boldsymbol{\nu} = 0 & \text{on } \Gamma, \end{cases}$$

where $\boldsymbol{\nu}$ is the exterior normal to W. They are given for $n \in \mathbb{N}$ by

$$u_n^{\pm}(x) = \begin{pmatrix} u_S^n(x_S) \\ \pm u_3^n(x_S) \end{pmatrix} e^{\pm i\beta_n x_3}, \qquad \begin{pmatrix} \boldsymbol{X}_n^{\pm}(x) \\ \boldsymbol{Y}_n^{\pm}(x) \end{pmatrix} = \begin{pmatrix} \pm\boldsymbol{\mathcal{X}}_n(x_S) \\ \boldsymbol{\mathcal{Y}}_n(x_S) \end{pmatrix} e^{\pm i\beta_n x_3},$$

where $\boldsymbol{\mathcal{X}}_n$ and $\boldsymbol{\mathcal{Y}}_n$ are the $\boldsymbol{X}$ and $\boldsymbol{Y}$ extensions of the transverse functions $(u_S^n,\, u_3^n)$. They satisfy $(\boldsymbol{\mathcal{X}}_n, \boldsymbol{\mathcal{Y}}_m)_\Sigma = \delta_{mn}$, where $(\cdot,\cdot)_\Sigma$ is a scalar product over $L^2(\Sigma)$ without complex conjugation. For a given frequency ω, β_n is real for only a finite number of guided modes, which are named propagating modes. The other ones are either inhomogeneous or evanescent. Because we only consider far fields, those modes will not be taken into account.
The assumption is then made that any elastic field, written in the $(\boldsymbol{X}, \boldsymbol{Y})$ variables, can be decomposed as follows:

$$\boldsymbol{X}|_\Sigma = \sum_n (\boldsymbol{X}, \boldsymbol{\mathcal{Y}}_n)_\Sigma \boldsymbol{\mathcal{X}}_n, \qquad \boldsymbol{Y}|_\Sigma = \sum_n (\boldsymbol{\mathcal{X}}_n, \boldsymbol{Y})_\Sigma \boldsymbol{\mathcal{Y}}_n.$$

IOP Conf. Series: Journal of Physics: Conf. Series **1131** (2018) 012010 doi:10.1088/1742-6596/1131/1/012010

We then consider a defect D inside the waveguide which lies between two sections $\Sigma_\pm$ and denote $\Omega = W\backslash\overline{D}$. The scattered field $\boldsymbol{u}_n^{s\pm}$ associated to the incident propagating mode $\boldsymbol{u}_n^\pm$ is solution of the following forward problem for a given frequency ω:

$$\begin{cases} \mathrm{div}\sigma(\boldsymbol{u}_n^{s\pm}) + \rho\omega^2\boldsymbol{u}_n^{s\pm} = 0 & \text{in } \Omega \\ \sigma(\boldsymbol{u}_n^{s\pm})\boldsymbol{\nu} = 0 & \text{on } \Gamma \\ \boldsymbol{u}_n^{s\pm} = -\boldsymbol{u}_n^\pm & \text{on } \partial D \\ (RC), & \end{cases}$$

with (RC) a radiation condition.
The data in this case are the components of the scattering matrix $\mathcal{S}$, namely the projections $S_{mn}^{+\pm}$, respectively $S_{mn}^{-\pm}$, along the $\boldsymbol{\mathcal{X}}_m$ on the two sections $\Sigma_\pm$ of the $\boldsymbol{Y}$ extensions $\boldsymbol{Y}_n^{s+}$ of the scattered fields $\boldsymbol{u}_n^{s+}$, respectively $\boldsymbol{Y}_n^{s-}$ of the scattered fields $\boldsymbol{u}_n^{s-}$. By only considering the propagating modes, the number of lines and columns of the scattering matrix is limited to $2P$, P being the number of propagating modes in each direction. The Linear Sampling Method consists in solving the following system for all sampling points $z = (z_S, z_3)$:

$$\begin{cases} \displaystyle\sum_{n=0}^{P-1} U_{mn}^{+-}h_n^- + U_{mn}^{--}h_n^+ = e^{i\beta_m(R+z_3)}\boldsymbol{\mathcal{Y}}_m(z_S)\cdot\boldsymbol{p}, \\ \displaystyle\sum_{n=0}^{P-1} U_{mn}^{++}h_n^- + U_{mn}^{-+}h_n^+ = e^{i\beta_m(R-z_3)}\boldsymbol{\mathcal{Y}}_m(z_S)\cdot\boldsymbol{p}, \end{cases} \qquad m = 0, \dots, P-1, \tag{1}$$

where

$$\mathcal{U} := \begin{pmatrix} U^{+-} & U^{--} \\ U^{++} & U^{-+} \end{pmatrix} = \begin{pmatrix} S^{+-} & S^{--} \\ S^{++} & S^{-+} \end{pmatrix} \begin{pmatrix} K & 0 \\ 0 & K \end{pmatrix} =: \mathcal{S}\mathcal{K},$$

K is the $P \times P$ diagonal matrix the components of which are $e^{i\beta_n R}/2i\beta_n$, $\pm R$ is respectively the x_3 coordinate of the section $\Sigma_\pm$ and $\boldsymbol{p}$ is a polarisation parameter. If, roughly speaking, a solution $H = (h^-, h^+)$, with $h^\pm = (h_1^\pm, \dots, h_P^\pm)$, is found, then $z \in D$ according to a classical result related to the LSM [2]. It is then possible to define a characteristic function of the defect over a sampling grid G, given by $\psi(z,\omega) = \|H\|_2^{-1}$, and so to obtain an image of the defect.
Let us remark that if every modes were considered in (1) (not only the propagating ones), the problem would be ill-posed as the near-field operator, from which the left-hand side of (1) is obtained, is compact [2]. The limitation to propagating modes can be viewed as a physical regularization of the near-field equation [13].

Back-scattering case The back-scattering case is a special case where only one side of the area to image, for example its left side, is accessible. It means that only data corresponding to sources and measurements on this side are available. In this case, the modal LSM system reduces to

$$\sum_{n=0}^{P-1} U_{mn}^{+-}h_n^- = e^{i\beta_m(R+z_3)}\boldsymbol{\mathcal{Y}}_m(z_S)\cdot\boldsymbol{p}.$$

2. The case of surface solicitations and measurements

The method shown above needs data within the waveguide, which is not realistic in the context of Non Destructive Evaluation. We write $\Gamma = \Gamma_0 \cup \Gamma_d$, d being the height of the waveguide and Γ_0 and Γ_d being respectively its lower and upper boundary. A family of source functions

$(\boldsymbol{g}_i^{\pm}(x))_{0\le i\le M-1} = (\boldsymbol{g}(x-x_i^{\pm}))_{0\le i\le M-1}$ is defined for an even and compactly supported function $\boldsymbol{g}$ defined on $\mathbb{R}$ with $x_i^{\pm} = \pm(R+i\delta)$. The diffraction problem satisfied by the total field $\boldsymbol{u}$ is:

$$\left\{\begin{array}{ccc} \mathrm{div}\sigma(\boldsymbol{u}) + \rho\omega^2\boldsymbol{u} = 0 & \text{in} & \Omega \\ \sigma(\boldsymbol{u})\boldsymbol{\nu} = \boldsymbol{g}_i^{\pm} & \text{on} & \Gamma_d \\ \sigma(\boldsymbol{u})\boldsymbol{\nu} = 0 & \text{on} & \Gamma_0 \\ \boldsymbol{u} = 0 & \text{on} & \partial D \\ (RC). & & \end{array}\right. \tag{2}$$

The corresponding scattered field $\boldsymbol{u}^s$ is $\boldsymbol{u} - \boldsymbol{u}^i$, where $\boldsymbol{u}^i$ solves the same problem (2) as $\boldsymbol{u}$ in W without the boundary condition $\boldsymbol{u} = 0$ on ∂D. To simplify the notations, a family of measurement functions $(\boldsymbol{f}_i^{\pm}(x))_{0\le i\le M-1} = (\boldsymbol{f}(x-x_i^{\pm}))_{0\le i\le M-1}$ is also defined for an even and compactly supported function $\boldsymbol{f}$ defined on $\mathbb{R}$ and for the same family of points $(x_i^{\pm})_{0\le i\le M-1}$. The assumption is also made that $\boldsymbol{f} = f\boldsymbol{e}_1$ and $\boldsymbol{g} = g\boldsymbol{e}_1$, the general case being given in [17]. Our data are the components of a measurement matrix $\mathcal{M}$ defined as follows:

$$\mathcal{M} = \begin{pmatrix} M^{+-} & M^{--} \\ M^{++} & M^{-+} \end{pmatrix},$$

where

$$M_{ij}^{-\pm} = \int_{\mathbb{R}} \boldsymbol{f}_i^{\pm}(x_3)\cdot\boldsymbol{u}^s(x_S,x_3)\,\mathrm{d}x_3$$

with $\boldsymbol{u}^s$ the solution of (2) for $\boldsymbol{g}_j^-$ and

$$M_{ij}^{+\pm} = \int_{\mathbb{R}} \boldsymbol{f}_i^{\pm}(x_3)\cdot\boldsymbol{u}^s(x_S,x_3)\,\mathrm{d}x_3$$

with $\boldsymbol{u}^s$ the solution of (2) for $\boldsymbol{g}_j^+$.
It can be shown [18] that the measurement matrix $\mathcal{M}$ is related to the LSM matrix $\mathcal{U}$ by the relationship

$$\mathcal{M} = -\mathcal{R}\mathcal{U}\mathcal{E}^T, \tag{3}$$

where $\mathcal{R}$ and $\mathcal{E}$ are the reception and emission matrices. Moreover, those matrices can be decomposed as

$$\mathcal{R} = \begin{pmatrix} R & 0 \\ 0 & R \end{pmatrix} \qquad \mathcal{E} = \begin{pmatrix} E & 0 \\ 0 & E, \end{pmatrix}$$

$$R = VF_gT, \qquad E = VF_fT,$$

where

- F_f and F_g are $P\times P$ diagonal matrices formed by the diagonal terms f_n and g_n respectively, where

$$f_n = \int_{\mathbb{R}} e^{i\beta_n z} f(z)\,\mathrm{d}z \qquad g_n = \int_{\mathbb{R}} e^{i\beta_n z} g(z)\,\mathrm{d}z;$$

- T is the $P\times P$ diagonal matrix formed by the diagonal terms $u_S^n(d)$;
- V is the $M\times P$ Vandermonde matrix given by

$$V_{mn} = e^{im\beta_n\delta}, \qquad m = 0,\dots,M-1, \quad n = 0,\dots,P-1.$$

The condition number of the emission $\mathcal{E}$ and reception $\mathcal{R}$ matrices can be rigorously analyzed and optimized [16, 18]: it strongly depends on the number $2M$ of sources and receivers and on the minimal distance δ between them. Inverting the system (3) enables us to compute $\mathcal{U}$ and then to apply the modal LSM as in section 1.
The above method can be extended to the 3D case by considering one or several lines of source and measurement points.

Back-scattering case Again in this case only one fourth of the data is available, which means that only one block of the measurement matrix is considered, namely M^{+-}. The equation to solve is then

$$M^{+-} = -RU^{+-}E^T,$$

the inversion of which enables us to solve the modal LSM in the back-scattering case.

3. The case of data in the time domain

In the time domain, we consider the following problem:

$$\begin{cases} \rho\partial_{tt}\boldsymbol{u} - \mathrm{div}\sigma(\boldsymbol{u}) = 0 & \text{in} \quad \Omega\times(0,+\infty) \\ \sigma(\boldsymbol{u})\boldsymbol{\nu} = \boldsymbol{g}_i^{\pm}(x)\chi(t) & \text{on} \quad \Gamma\times(0,+\infty) \\ \boldsymbol{u} = 0 & \text{on} \quad \partial D\times(0,+\infty) \\ \boldsymbol{u} = \partial_t\boldsymbol{u} = 0 & \text{on} \quad \Omega\times\{0\}. \end{cases}$$

The data consist of the corresponding scattered fields measured at the same points as before in the time interval $(0,+\infty)$. Here, the function $\chi(t)$ is suitably chosen so that the frequencies for which the group velocity vanishes are avoided. By applying a Fourier transform to our data we recover, up to a $\widehat{\chi}(\omega)$ factor, $\widehat{\chi}$ being the Fourier transform of χ, the previous system (2) at a given frequency ω. We now have multi-frequencies data, which allows us to image the defect with a better accuracy than in the frequency domain. More precisely, as explained in [16], the following combination is chosen:

$$\Psi(z) = \left(\int_{\omega_-}^{\omega_+} \frac{\max_{z'\in G}|\psi(z',\omega)|^2}{|\psi(z,\omega)|^2}\,\mathrm{d}\omega\right)^{-1/2},$$

where ω_- and ω_+ are such that $\mathrm{supp}\widehat{\chi}\subset(\omega_-,\omega_+)$ and G is the sampling grid.
In figure 1, the reconstruction is done using multiple frequencies which correspond to a number of propagating modes P ranging from 8 to 14. More precisely, the height of the waveguide is $d = 1$ mm and the material constants are $\lambda = 121$ GPa, $\mu = 80$ GPa and $\rho = 7900$ kg.m^{-3}. No noise was added in this case as it has already been seen in the acoustic case [16] that the method is robust regarding to noise. Furthermore, some noise is naturally present in the experimental data considered in the next section.

4. Application to experimental data

The method has been tested for data measured on a steel plate (see figure 2), the parameters of which are $\rho = 7926$ kg.m^{-3}, $d = 12$ mm, $\lambda = 110.75$ GPa and $\mu = 82.435$ GPa.

In order to use a 2D waveguide model for the plate, it is needed that the defects and sensor are considered as invariant according to one direction, namely the x_2 direction. Therefore, the plate can be separated in 3 different areas:

- a zone without defect allowing the measurement of the incident field;
- a zone containing a cylindrical notch of rectangular section, its dimension being 0.25 mm in the x_3 direction and 2 mm in the x_1 one;
- a zone containing a vertically centered cylindrical hole of circular section, its diameter being of 2 mm.

A single linear multi-element piezoelectric sensor was used to obtain the data. More precisely, each of the transducer was successively used as a source while all of them were used as measurement points. Its main characteristics are $\delta = 0.8$ mm (element width of 0.55 mm

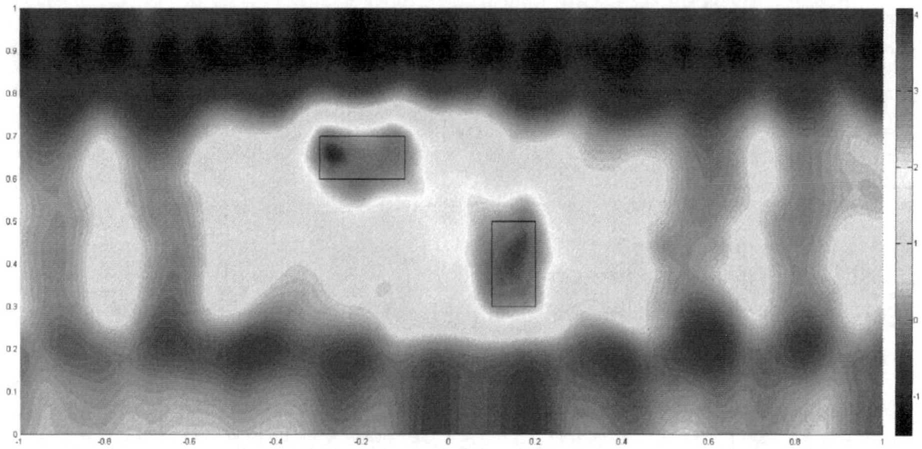

Figure 1: Defect identification from artificial data, $M = 42$ (log scale, axes unit: mm)

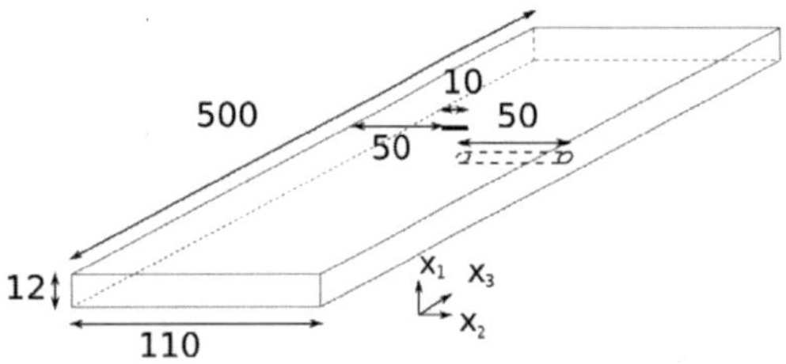

(a) Scheme of the plate (unit: mm)

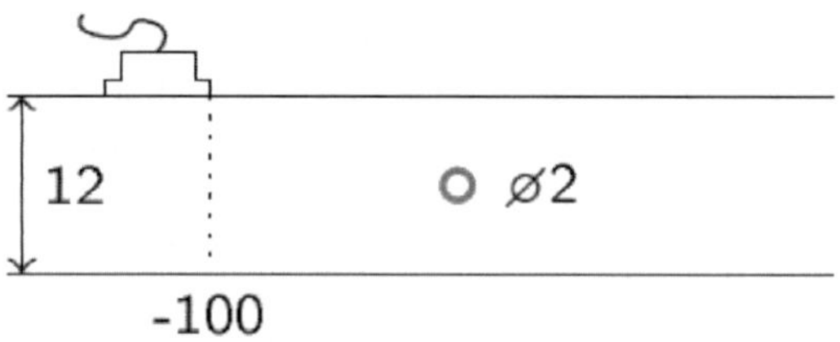

(b) Setup in the case of the cylindrical hole

Figure 2: Experimental setup

and inter-element space of 0.25 mm), M =128 and its central frequency is of 2 MHz, which means that $P \approx 20$. Its element's bandwidth at -6 dB is of 1.2 MHz and the sampling frequency used is of 10 MHz. Its elements have a width of 18 mm enabling us to use the 2D approximation. The fact that only one sensor was used implies that the case under consideration is the back-scattering one.
In the following images, D will denote the distance between the sensor and the defect that is being imaged according to the x_3 direction. In each of the images, the sensor is on the upper left corner, outside of the image (see Figure 2 (b) for an example). In Figure 3, the cylindrical hole is imaged for various values of D. It can be seen that the defect is well retrieved for every value of D, even though the resolution slightly decreases with the distance. It can be explained by some attenuation that is not taken into account in our model. The resolution of the defect is good according to the section and degraded a little along the axis, which is expected in the back-scattering configuration [13]. The same results are obtained for the cylindrical notch (see Figure 4).

Conclusion and perspectives

The method presented here allow the imaging of defects inside 2D elastic waveguides, that is a section of a plate, from surface data. Further testings need to be done regarding its application to other kind of materials, for example multi-layered waveguides, and three-dimensional waveguides such as rails or pipes.

IOP Conf. Series: Journal of Physics: Conf. Series **1131** (2018) 012010 doi:10.1088/1742-6596/1131/1/012010

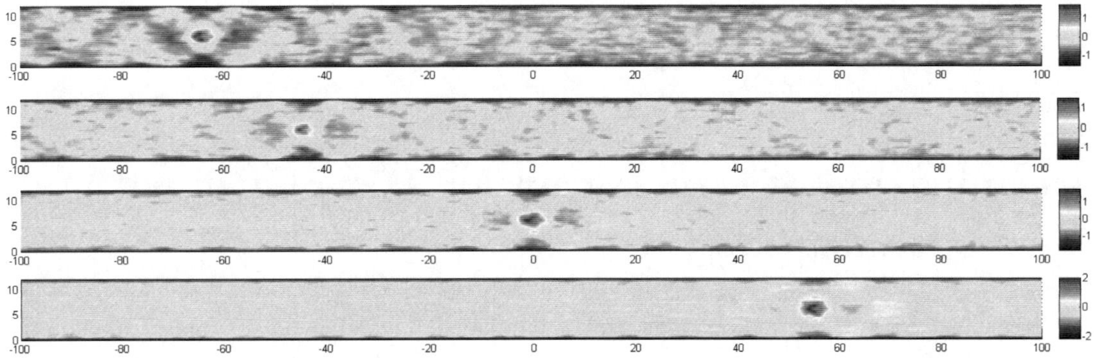

Figure 3: Identification of the cylindrical hole for $D = 35$, $D = 55$, $D = 95$ and $D = 150$ (log scale, axes unit: mm).

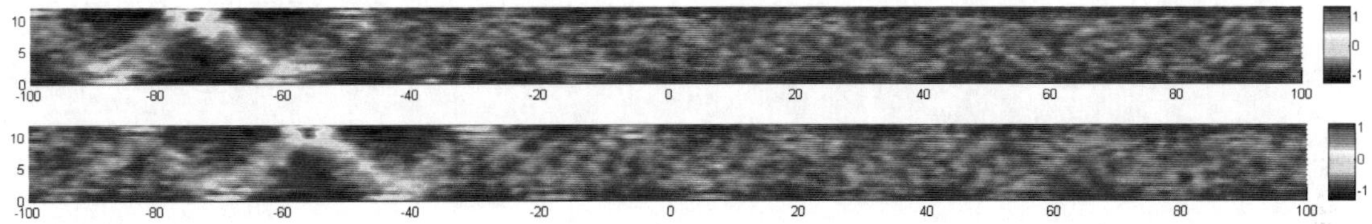

Figure 4: Identification of the cylindrical notch for $D = 25$ and $D = 45$ (log scale, axes unit: mm).

References

[1] Colton D and Kirsch A 1996 *Inverse Problems* **12** 383-393
[2] Bourgeois L, Le Louër F and Lunéville E 2011 *Inverse Problems* **27**
[3] Huthwaite P 2014 *Proceedings of the Royal Society of London* **470**
[4] Fletcher S, Lowe M, Ratassepp M and Brett C 2012 *Journal of Nondestructive Evaluation* **31** 56-64
[5] Rao J, Ratassepp M, Fan Z 2016 *IEEE Transactions on Ultrasonics, Ferroelectrics, and Frequency Control* **63** 737-745
[6] Rodriguez S, Deschamps M, Castaings M and Ducasse E 2014 *Ultrasonics* **54** 1880-1890
[7] Charalambopoulos A, Gintides D, Kiriaki K 2003 *Inverse Problems* **19** 549-561
[8] Haddar H and Monk P 2002 *Inverse Problems* **18** 891-906
[9] Borcea L and Nguyen D 2016 *Inverse Probl. Imaging* **10** 915-941
[10] Chen Q, Haddar H, Lechleiter A, Monk P 2010 *Inverse Problems* **26**
[11] Monk P and Selgas V 2016 *Inverse Problems* **32**
[12] Catapano I, Crocco L, D'Urso M and Isernia T 2009 *Inverse Problems* **25**
[13] Bourgeois L and Lunéville E 2008 *Inverse Problems* **24**
[14] Bourgeois L and Lunéville E 2013 *Inverse Problems* **29**
[15] Bourgeois L and Fliss S 2014 *Inverse Problems* **30**
[16] Baronian V, Bourgeois L and Recoquillay A 2016 *Wave Motion* **66** 66-87
[17] Recoquillay A 2018 *PhD thesis*
[18] Baronian V, Bourgeois L, Chapuis B and Recoquillay A 2018 *Inverse Problems*

IOP Conf. Series: Journal of Physics: Conf. Series **1131** (2018) 012011 doi:10.1088/1742-6596/1131/1/012011

Determination of blur kernel for HR-pQCT with bilevel optimization

B.Sixou[1], **Y.Li**[1], **F.Peyrin**[1]

[1] CREATIS, INSA-Lyon; Inserm, U1044; CNRS UMR 5220; Université Lyon 1; Université de Lyon; 69621, Lyon, France

E-mail: bruno.sixou@insa-lyon.fr

Abstract. In this work, we determine the blur kernel of HR-pQCT scanner with bilevel optimization. The method is compared with an optimization of the kernel based on the L_2 distance followed by regularization with Total Variation. An improvement is obtained for the reconstructed image.

1. Introduction

X-ray CT techniques are well adapted to image bone structure. However, imaging bone micro architecture *in vivo* remains challenging because it is difficult to achieve a high spatial resolution (order of 100μ m) with a low X-ray dose. High Resolution peripheral Quantitative CT (HR-pQCT) providing images with spatial resolution of about 100-150μm have been developed recently[1]. However, the segmentation and the extraction of bone micro-structure parameters remains an issue since the size of the structures is of the same order than the spatial resolution. To improve the spatial resolution, super resolution techniques may be applied. However, the HR-pQCT blurring kernel involved in the image formation process is unknown and has to be determined. The aim of this work is to investigate the degradation kernel of HR-pQCT for given high resolution and low resolution images. We consider the image acquired from parallel-beam synchroton micro-CT as the high resolution image[2], and the image generated by HR-pQCT as the low resolution image. In a previous work, we have used L_2 norm and mutual information minimization without any regularization. The kernels obtained with these approaches are then used for a classical Total Variation regularization with real data to solve the super-resolution problem. [3, 4, 5].

We determine here the blur kernel of HR-pQCT with a bilevel optimization method. Bilevel approaches for learning variational models have been much developed recently in image processing and inverse problems[6, 7, 8, 9]. For example, bilevel optimization for finite dimensional Markov random field models is presented in [10, 11] A bilevel approach for blind deconvolution is discussed in [6]. In this work, the bilevel approach is used and formulated in a discrete setting. The regularization functional includes a non-differentiable TV term but a smoothed approximation is considered for numerical computation.

This paper is organized as follows. After the introduction, the HR-pQCT inverse problem considered is summarized and it is formulated as a bilevel optimization problem. The method to solve the optimization problem based on the adjoint method is presented. Numerical results are diplayed before concluding remarks.

IOP Conf. Series: Journal of Physics: Conf. Series **1131** (2018) 012011 doi:10.1088/1742-6596/1131/1/012011

2. Bilevel framework for the HR-pQCT imaging problem

2.1. The image problem formulation

The reconstruction of a 3D image with an improved resolution from a single low-resolution image can be expressed as follows. We assume the low and high resolution images are defined on a bounded region Ω. The low-resolution image is obtained from a high-resolution image with blur, down sampling and noise. The forward problem can thus be written as:

$$g = U(h * f) + \epsilon \tag{1}$$

where $g \in \mathbb{R}^N$ denotes the N-voxels 3D low-resolution noisy image, and $f \in \mathbb{R}^{N'}$ denotes an $N' = N \times p^3$-voxels high-resolution image with super-resolution factor $p = 2$ in each dimension, ϵ is the additive noise component, U is the undersampling operator and h the convolution kernel accounting for the blurring of the images. The down sampling operator is described as $g_{[i,j,k]} = f_{[2i-1,2j-1,2k-1]}$ for all indexes i, j, k. Other voxels of f are discarded during the sampling process. In this work, we do not try to optimize this operator U.

2.2. Bilevel optimization formulation

In this work, we investigate a smoothed bilevel optimization approach. We focus on the following upper-level optimization problem:

$$\min_h \hat{J}(h) = \|f(h) - f_{true}\|^2 \tag{2}$$

where f_{true} is the ground truth image obtained with high resolution CT. The image $f(h)$ is defined as the solution of the smoothed lower level variational problem:

$$f(h) = \arg\min_f \|U(h * f) - g\|^2 + \alpha\phi_\gamma(Df) \tag{3}$$

for a regularization parameter $\alpha > 0$. In this formula, D denotes the discrete gradient and ϕ_γ is a smoothing function approximating the L_1 norm.

We assume that the lower level optimization problem admits an unique solution.

With this regularization approach of the L_1 norm, the first order optimality condition can be easily derived [7]. In the following, we consider the $C^2(\mathbb{R}, \mathbb{R})$ Huber smoothing function:

$$\phi_\gamma(t) = \begin{cases} \frac{3\gamma}{8} + \frac{3}{4\gamma}|t|^2 - \frac{1}{8\gamma^3}|t|^4 & |t| < \gamma \\ |t| & |t| \geq \gamma \end{cases} \tag{4}$$

We have also

$$\phi'_\gamma(t) = \begin{cases} \frac{-t^3}{2\gamma^3} + \frac{3t}{2\gamma} & |t| < \gamma \\ sgn(t) & |t| \geq \gamma \end{cases} \tag{5}$$

$$\phi''_\gamma(t) = \begin{cases} \frac{-3t^2}{2\gamma^3} + \frac{3}{2\gamma} & |t| < \gamma \\ 0 & |t| \geq \gamma \end{cases} \tag{6}$$

For a vector $a = (a_1, ...a_N)^t$, $\phi_\gamma(a) = (\phi_\gamma(a_1), ...\phi_\gamma(a_N))^t \in \mathbb{R}^N$

2.3. Adjoint method.

The necessary and sufficient optimality condition can be written with $E : \mathbb{R}^{N'} \times \mathbb{R}^{N'} \to \mathbb{R}^{N'}$[7]:

$$E(f, h) = h_- * h * f - h_- * U^t * g + \alpha D^t \phi'_\gamma(Df) = 0 \tag{7}$$

where h_- is the kernel defined by $h_-(x) = h(-x)$ for $x \in \Omega$. We will assume that the kernel is symmetric and $h_-(x) = h(-x) = h(x)$ for $x \in \Omega$.
In the following, we will admit that the constraint operator E is differentiable with smooth derivatives. Let $v = (f, h)$, the first derivative of the mapping f in the direction δh is :

$$f'(h)\delta h = -E_f^{-1}(v)E_h(v)\delta h \tag{8}$$

where E_f and E_h denote the derivatives of E with respect to f and h respectively. By the chain rule, we obtain the gradient of $\hat{J}$ at a control h :

$$\nabla \hat{J}(h) = J_h(h) - E_h^*(v)E_f^{-*}(v)J_f(h) \tag{9}$$

where E_h^* and E_f^{-*} denotes the adjoints of the operators.

In the framework of the adjoint method, we introduce the variable p defined as

$$p = -E_f^{-*}(v)J_f(v) \tag{10}$$

and the gradient can be rewritten:

$$\nabla \hat{J}^k(h) = J_h(h) + E_h^*(v)p \tag{11}$$

For every $v = (f, h) \in \mathbb{R}^{N'} \times \mathbb{R}^{N'}$, $E_f(f, h) : \mathbb{R}^{N'} \to \mathbb{R}^{N'}$is given by:

$$E_f(f, h)w = h_- * h * w + \alpha D^t \phi''_\gamma(Df) \otimes Dw \tag{12}$$

where $a \otimes b = (a_1 b_1, \ldots\ldots, a_n b_n)$. Let H the linear operator associated to the convolution by the operator h. The function ϕ''_γ is positive and thus, assuming that $KerH = \{0\}$, we obtain that E_f is an homeomorphism.

With $J_f = f - f_{true}$, p is the solution of the adjoint equation:

$$E_f^*(f, h)p = h_- * h * p + \alpha D^t \phi''_\gamma(Df) \otimes Dp = -(f - f_{true}) \tag{13}$$

We have also to compute the Fréchet derivative of the constraint with respect to h and its adjoint. The operator $E_h(f, h) : \mathbb{R}^{N'} \to \mathbb{R}^{N'}$, is given by

$$E_h(\delta h) = \delta h_- * (h * f - g) + h_- * \delta h * f \tag{14}$$

Its adjoint $E_h(f, h)^* : \mathbb{R}^{N'} \to \mathbb{R}^{N'}$ is given by

$$E_h(f, h)^* u = u * (h * f - g)_- + f_- * h * u \tag{15}$$

The evaluation of the gradient $\nabla \hat{J}^k(h_k)$ at the point h_k can be summarized with the following algorithm:
1) Solve the low level optimality equation to obtain f_k from the current approximation (h_k) with Eq.7.
2) Solve the backward adjoint equation to obtain the adjoint variable p_k, with Eq.13.
3) Return the gradient with respect to h with Eq.11.

IOP Conf. Series: Journal of Physics: Conf. Series **1131** (2018) 012011 doi:10.1088/1742-6596/1131/1/012011

3. Numerical simulation

3.1. Implementation details

In order to optimize the blurring kernel, we use a entire dabase of pairs of 3D ground truth and corresponding degraged images of trabecular bone[12]. The imaging data were collected for a custom-built bone structure phantom. The phantom was comprised of cadaveric distal tibia sections (1cm thick) embedded in an 8 cm diameter cylinder comprised of polymethylmethacylate (PMMA) and polyethylene resin. The phantom was scanned using typical in vico protocol settings by HR-PQCT (XtremeCt, Scanco Medical AG), which generates reconstruction with 82 μ m isotropic voxels [1] . The reference scan data was obtained from parallel-beam synchrotron micro-CT [2] images which provided high resolution reconstructions, with 24 μm isotropic voxels, After registration with B-spline interpolation, the voxel size was equal to 41 μm. In this work, we consider that images that are perfectly aligned to each others such that only a deconvolution and upscaling is necessary.

The voxel size of the low and high resolution images are 82 μm and 41 μm respectively. The size of the HR-pQCT μCT images is $420 \times 510 \times 110$, and $840 \times 1020 \times 220$ respectively. For the training step two tibia samples are used in which 50 crops of size $100 \times 100 \times 100$ and $200 \times 200 \times 200$ are cut from the images. At the end of the training process, one test crop is selected. Some sections of the high and low resolution images for the test crop are displayed in Figure 1. The second approach is detailed in [3]. The optimization of the kernel is performed with the minimization of the square of the L_2 norm $\|U(h * f) - g\|^2$. This is followed by a TV super resolution restoration with ADMM (Alternate Direction of Multipliers Method) with optimized parameters. This second method is similar to the traditional blind deconvolution techniques with an unknown convolution kernel [13]. The Otsu segmentation method is applied to calculate DICE, BV/TV (bone volume/total volume) for the two approaches in order to evaluate the quality of the reconstruction.The DICE is defined as $\frac{|f_b \cap f_b^*|}{|f_b|+|f_b^*|}$ where f_b and f_b^* are binary images of f and f^* respectively. $|f_b \cap f_b^*|$ counts the number of elements where the two binary images overlap. The BV/TV is the percentage of the bone volume (BV) with respect to total volume (TV) in the image. BV is simply calculated by counting the number of 1 in the segmented image.

3.2. Results and discussion

The reconstructed images with the bilevel approach is displayed in Figure 2, together with the estimated kernel. At the end of the training process, the relative norm change between the initial guess h_0 and the estimated kernel $\hat{h}$, $\frac{\|\hat{h}-h_0\|}{\|h_0\|}$ is 0.08. Table 1 summarizes the DICE, BV/TV for the two methods. As explained above, the DICE is a measure of the distance between the binary high resolution image and the binary reconstructed image. The bilevel method improves this error measure. For the high resolution image, the BV/TV value is 0.25. This structural parameter is also improved with the bilevel approach.

Promising results are obtained with the bilevel optimization scheme and the DICE and BV/TV are slightly improved. It should be noted that the regularization parameter is fixed during the algorithm and that it could be also optimized in a more general framework.

4. Conclusion

In this paper, we have used a bilevel approach to determine the kernel of HR-pQCT. The upper optimization problem is the L_2 norm distance between the image obtained with a smoothed TV regularization and the ground truth. This method is compared with a L_2 data term norm minimization followed by a TV regularization on bone micro-structure images. Promising results are obtained with the bilevel optimization. In future studies, the optimization of the

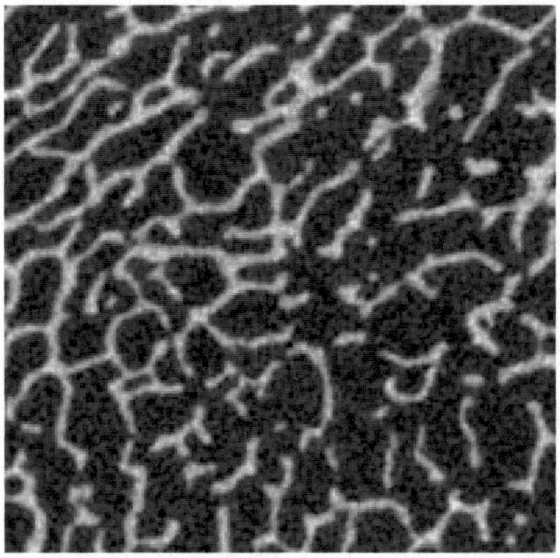

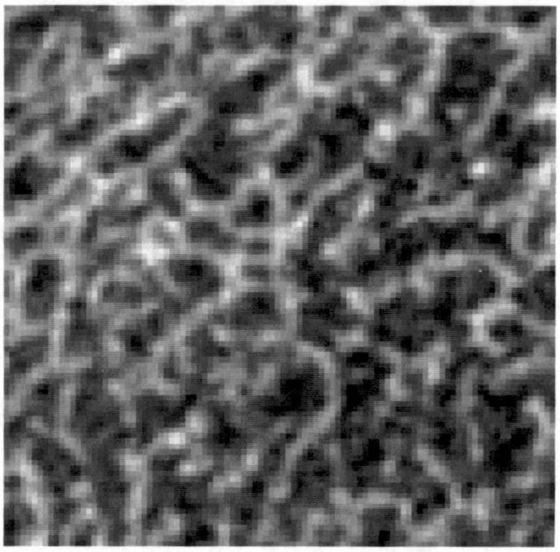

(a) (b)

Figure 1: (a) High resolution image. (b) Low resolution image.

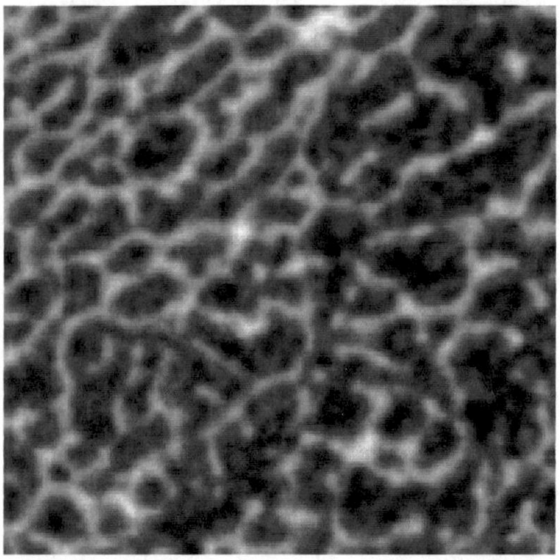

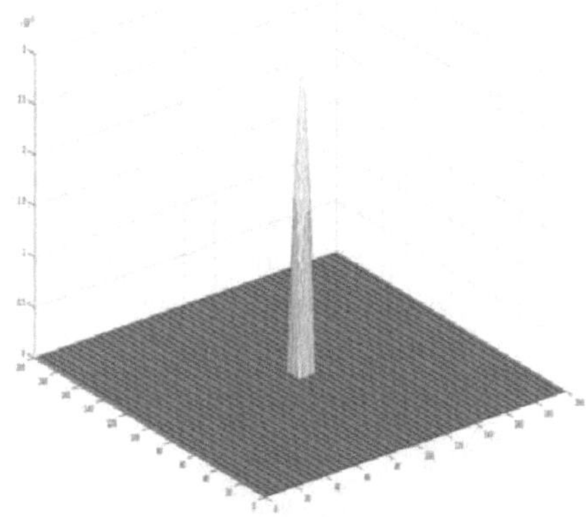

(a) (b)

Figure 2: (a) Reconstructed image with bilevel approach. (b) Kernel obtained with bilevel optimization

Table 1: DICE, BV/TV for the two approaches. For the high resolution μCT image, the BV/TV is 0.25.

kernel	DICE	BV/TV
Bilevel optimization	0.78	0.32
$L_2 + TV$	0.74	0.4

regularization parameter will be inserted in the bilevel optimization method with a more general functional.

IOP Conf. Series: Journal of Physics: Conf. Series **1131** (2018) 012011 doi:10.1088/1742-6596/1131/1/012011

References

[1] S.Boutroy, M.L.Bouxsein, F.Munoz, P.Delmas *In vivo assessment of trabecular bone microarchitecture by high-resoltuion peripheral quantitative computed tomography,* Inverse Problems and Imaging, **90** (2005) 6508-6515 .

[2] M.Salome, F.Peyrin, P.Cloetens, C.Odet, A.M.Laval-Jeantet, J.Baruchel and Per Spanne *A synchrotron radiation microtomography system for the analysis of trabecular bone samples,* Medical Physics, **26** (1999) 2194-2204.

[3] Y.Li, B.Sixou and F.Peyrin *Estimation of the blurring kernel in experimental HR-pQCT images based on mutual information,* European Signal Processing Conference, Eusipco, (2017) 2086-2090.

[4] A.Toma, L.Denis, B.Sixou, J.B.Pialat and F.Peyrin *Total variation super-resolution for 3D trabecular bone micro-structure segmentation,* European Signal Processing Conference, Eusipco, (2014) 2220-2224.

[5] A.Toma, B.Sixou and F.Peyrin *Iterative choice of the optimal regularization parameter in TV image restoration* , Inverse Problems and Imaging, **9** (2014) 860-867.

[6] M.Hintermuller and T.Wu "Bilevel optimization for calibrating point spread functions in blind deconvolution", *Inverse Problems in Imaging,* vol.9, 1139-1169 , 2015.

[7] K.Kunisch and T.Pock "A bilevel optimization approach for parameter learning in variational models", *SIAM J.Imaging Sci.*, vol.6, 938-983 , 2013.

[8] J.C.De Los, Reyes, C.B.Schonlieb, T.Valkonen "The structure of optimal parameters for image restoration problems", *Journal of Mathematical Analysis and Applications,* vol.434, 464-500 , 2016.

[9] Y.Chen, T.Pock and H.Bischof "Learning of l_1 based analysis and syntheis sparsity priors using bi-level optimization", *Workshop on Analysis Operator Learning vs.Dictionary learning,* NIPS 2012.

[10] Y.Chen, R.Ranftl and T.Pock *Insights into analysis operator learning: from patch-based sparse modesl to higher-order MRFs,* IEEE Transactions on Image Processing, **23** (2014) 1060-1072.

[11] S.Roth and M.J.Black *Fields of experts: a framework for learning image priors,* Computer Vision and Pattern Recognition, **2** (2005) 860-867.

[12] A.J.Burghardt, J.B.Pialat, G.J.Kazakia, S.Boutroy, K.Engelke, J.M.Patsch, A.Valentinitsch, D.Liu, E.Szabo, C.E.Bogado *Multicenter precision of cortical and trabecular bone quality measures assessed by high-resoution peripheral quantitative computed tomography,* Journal of Bone and Mineral Research, **28** (2013) 524-536.

[13] M.Burger and O.Scherzer *Regularizatin methods for blind deconvolution and blind source separatio problems,* Math. Control Signal Syst., **14** (2001) 358-383.

IOP Conf. Series: Journal of Physics: Conf. Series **1131** (2018) 012012 doi:10.1088/1742-6596/1131/1/012012

Advanced statistical learning method for multi-physics NDT-NDE

S Ahmed[1], P Calmon[1], R Miorelli[1], C Reboud[1] and A Massa[2,3]
[1]CEA LIST, Centre de Saclay, F- 91191 Gif-sur-Yvette, France
[2]ELEDIA Research Center (ELEDIA@UniTN - University of Trento)
Via Sommarive 9, I-38123 Trento, Italy
[3]ELEDIA Research Center (ELEDIA@L2S - UMR8506)
3 rue Joliot-Curie 91192 Gif-sur-Yvette, France

E-mail: roberto.miorelli@cea.fr

Abstract. This work presents an innovative multi-physics (MP) Learning-by-Examples (LBE) inversion methodology for real-time non-destructive testing (NDT). Eddy Current Testing (ECT) and Ultrasonic Testing (UT) data are effectively combined to deal with the localization and characterization of a crack inside a conductive structure. An adaptive sampling strategy is applied on ECT-UT data in order to build an optimal (i.e., having minimum cardinality and highly informative) training set. Support vector regression (SVR) is exploited to obtain a computationally-efficient and accurate surrogate model of the inverse operator and, subsequently, to perform real-time inversions on previously-unseen measurements provided by simulations. The robustness of the proposed MP-LBE approach is numerically assessed in presence of synthetic noisy test set and compared to single-physic (i.e., ECT or UT) inversion.

1. Introduction

Real time accurate inversion solution becomes the main priority in non-destructive testing and evaluation (NDT-NDE) applications. Among different iterative [1, 2, 3] and non-iterative [4, 5, 6, 7] inversion solutions, Learning by examples (LBE) strategy is getting more attention for having quasi-real time inversion capabilities. In this work, LBE has been adopted for a NDE problem where a narrow crack is occurred around a fastener (e.g., bore hole) within an inspected medium [8]. This is an important problem for the aging aircraft NDE community and Eddy Current Testing (ECT) is widely applied while the structure under test (SUT) thickness is thin. However, the penetration depth of the induced currents is limited by the skin depth. This makes the detection and resolution of defects more difficult as the depth increases. Whereas ultrasound testing (UT) NDT inspection is suitable for high resolution, but the inspection is affected by the surface roughness of the inspected medium [9]. That means, each of these NDT methods has some pros and cons according to their own physics. Moreover, for the mentioned problem at hand, ECT signal is mostly affected for the presence of fastener. Due to the significant probe impedance variation, the area of the fastener is acting as a circular defect within the inspected medium. The impedance variation due to the fastener is much stronger than narrow crack, thus, when the crack is placed deeper inside the SUT, the ECT signals contribution due to the presence of narrow crack becomes weaker. Conversely, UT signals are stronger for subsurface crack compare to the crack placed at the top surface of the SUT. As a consequence, different impacts on the crack characterization and localization performance are expected based on ECT and UT methods. Thus, multi-physics (MP) data fusion (ECT-UT) has been applied to maximize the inversion performance for crack characterization and localization.

In general, LBE is a two phases approach. During the preliminary phase (so called offline phase), a fast and accurate inverse/trained model is built based on a training set made of input-output (I/O) pairs by learning algorithm. The developed (trained) model from offline phase is then used to predict the output associated to an unknown test sample during the second phase (online phase). Within the framework of LBE, an adaptive sampling strategy combining Partial Lest Square (PLS) [10] feature extraction and modified version of output space filling (OSF) [11] (i.e., PLS-OSF sampling [4]) has been adopted for obtaining optimal training sets by both ECT and UT methods separately. An updated version of PLS-OSF sampling algorithm has also been illustrated for dealing with ECT-UT data. Support vector regression (SVR) [12] is used to obtain accurate training model and perform real time inversion. Finally, the performance of the MP-LBE inversion schema for crack characterization and localization is compared to single-physic (i.e., ECT and UT) inversion on noisy data.

2. Mathematical formulation of forward and inverse problem

Let us consider a homogeneous plate made by aluminium 2024 alloy of thickness 6 mm, density 2.77 g.cm^{-3}, has been investigated by both ECT and UT NDT methods. The plate consists of a fastener (bore hole) of radius 3.75 mm and 6.00 mm height. The plate is affected by a single notch (e.g., narrow crack) of volume Ω having fixed width 0.01 mm and height 2 mm (Fig. 1) which is attached with the fastener. The crack is characterized by total $Q = 3$ descriptors of length (l_c), ligament (δ_c) and angular distance (φ_c) (i.e., $\underline{p} = (l_c, \delta_c, \varphi_c)$).

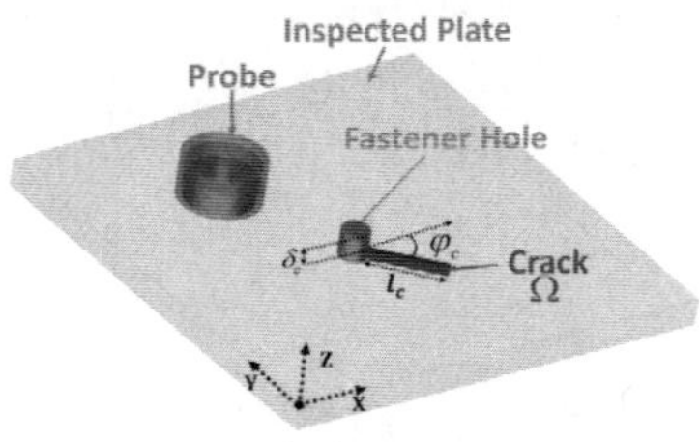

(*a*)

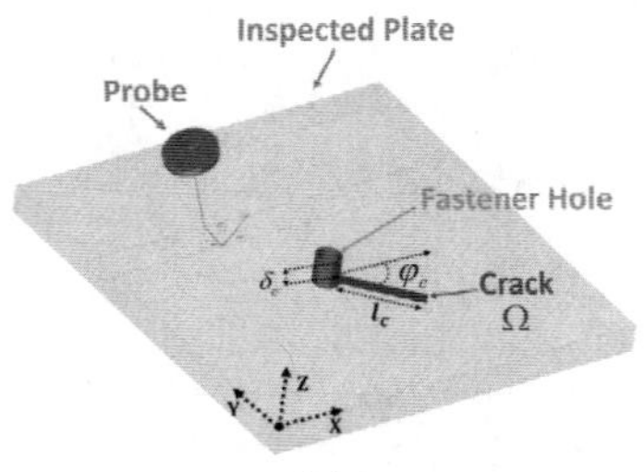

(*b*)

Figure 1. Examples of studied (*a*) ECT and (*b*) UT configuration

2.1. ECT treatment

The plate is inspected by a single coil working in absolute mode of frequency 1 kHz with lift off 1 mm. The coil impedance variation due to presence of the crack measured at the $k-$th ($k = 1, ..., K$) scanning position with respect to the flawless region is given by [13].

$$\chi_k^{ECT} = \frac{1}{I^2}\int_{\Omega} \mathbf{E}^{inc}(\mathbf{r}|\mathbf{r}_k).\rho(\mathbf{r}|\mathbf{r}_k)d\mathbf{r} \tag{1}$$

I is the current flowing inside the coil while $\mathbf{E}^{inc}(\mathbf{r}|\mathbf{r}_k)$ is the incident field generated at position $\mathbf{r}$ in the unflawed plate ($\mathbf{r}_k = (x_k, y_k)$ represents the k-th coil position within the plate). $\rho(\mathbf{r}|\mathbf{r}_k)$ is the unknown induced current dipole density, which models the presence of the crack and is related to the total field, $\mathbf{E}^{tot}(\mathbf{r}|\mathbf{r}_k)$ that can be expresses by $\rho(\mathbf{r}|\mathbf{r}_k) = [\sigma(\mathbf{r}) - \sigma]\mathbf{E}^{tot}(\mathbf{r}|\mathbf{r}_k)$. CIVA simulator [14] has been used as a forward operator $\Phi^{ECT}\{.\}$ in order to generate ECT coil signals $\underline{\chi}^{ECT}$. Due to the complex nature of the ECT signals, ECT signals are represented by $\underline{\chi}^{ECT} = \left\{\left(\Re\left\{\chi_k^{ECT}\right\};\Im\left\{\chi_k^{ECT}\right\}\right); k = 1, ..., K\right\}$ of $F^{ECT} = 2K$ ECT features.

2.2. Ultrasound testing treatment

The plate has been investigated by a ray probe by using water coupling medium (i.e., density 1 g.cm^{-3}). The probe is acting both for transmitting and receiving UT signals. More details of the treated problem and probe definition are available in [15]. CIVA uses a hybrid model known as Physical Theory of Diffraction (PTD), based on Kirchhoff approximation and high-frequency Geometrical Theory of Diffraction (GTD) for generating diffraction/scattering waves from planner-like defects. The scattered field, for the PTD model can be expressed by [16]

$$\mathbf{u}^{Scat(PTD)}(\mathbf{r}) = \mathbf{u}^{Rayleigh}(\mathbf{r}) + \mathbf{u}^{Kir}(\mathbf{r}) + \sum_{\beta} \left[\left(D_{\beta}^{\alpha(GTD)}(\mathbf{r}) - D_{\beta}^{\alpha(KA)}(\mathbf{r}) \right) \frac{e^{i\lambda_{\beta} S_{\beta}}}{\sqrt{\lambda_{\beta} L_{\beta}}} \mathbf{e}_{\beta}(\mathbf{r}) \right]. \quad (2)$$

where, $\alpha = L, TV, or\, TH$ (Longitudinal, Transverse Vertical or Transverse Horizontal, respectively) incident type wave vector and $\beta = L, TV, or\, TH$ is the scatter type wave vector. S_{β} is the distance between the diffraction point $\mathbf{r}_{\beta}^{\alpha}$ and the observation point $\mathbf{r}$. L_{β} is a parameter distance. $D_{\beta}^{\alpha(GTD)}$ is the GTD diffraction coefficient and $D_{\beta}^{\alpha(KA)}$ is the Kirchhoff edge diffraction coefficient. $\mathbf{u}^{Kir}$ is the displacement scattered field at the observation $\mathbf{r}$, and the Rayleigh field $\mathbf{u}^{Rayleigh}(\mathbf{r})$ comprises the surface waves. The reflections/scatters wave have been collected through C-Scan (e.g., maximum ray amplitude available at each inspection point) and represented by $\underline{\chi}^{UT}$. CIVA simulator [14] has been utilized as a forward operator $\Phi^{UT}\{.\}$ in order to generate UT data. Unlike, ECT signals, UT signals contain only the real data, hence UT signals are represented by $\underline{\chi}^{UT} = \left\{\chi_k^{UT}; k = 1, ..., K\right\}$ of $F^{UT} = K$ UT features.

2.3. Data fusion using ECT and UT

In this case, ECT signals and UT signals are generated separately by their own forward solver (i.e., $\Phi^{ECT}\{.\}$ and $\Phi^{UT}\{.\}$) and both of these data sets are fused by concatenating ECT and UT data. The obtained fused ECT-UT data are represented by $\underline{\chi}^{ECT-UT} = \left\{\underline{\chi}^{ECT}; \underline{\chi}^{UT}\right\}$ of $F^{ECT-UT} = 3K$ ECT-UT features.

2.4. Adaptive sampling through feature extraction

The main goal of the adaptive sampling (i.e., PLS-OSF) is to apply PLS feature extraction for reducing the dimension of the actual features (e.g., F^{ECT}, F^{UT} and F^{ECT-UT}) by projecting into extracted feature space. After-which, adaptive sampling is performed directly in the extracted feature space to build suitable I/O pairs for building optimal training model by using lowest number of training samples during *offline* phase. Though, PLS-OSF [4] has been directly applied for ECT and UT data separately, a modified version of PLS-OSF is needed for dealing with ECT-UT data. The following steps describe the updated PLS-OSF sampling strategy.

i **Initialization**- Generate N_0 number of initial samples by uniform GRID (i.e., full factorial grid) sampling approach. A matrix of defect parameters $\underline{\underline{p}} = (\underline{p}^{(n)}; n = 1, ..., N_0)$ having $(N_0 \times Q)$ dimension is formed, where $\underline{p}^{(n)}$ is the n-th row of $\underline{\underline{p}}$. By using $\Phi^{ECT}\{.\}$ and $\Phi^{UT}\{.\}$ generate ECT and UT data respectively, and fill the $\left(N_0 \times F^{ECT-UT}\right)$ feature matrix $\underline{\underline{\chi}}^{ECT-UT} = \left\{\left(\underline{\chi}^{ECT-UT}\right)^{(n)}; n = 1, ..., N_0\right\}$.

ii **PLS Feature Extraction**- In this step, the $F^{ECT-UT} = 3K$ dimensional ECT-UT data are reduced to J number of extracted features where, $J << F^{ECT-UT}$. A $\left(N_0 \times F^{ECT-UT}\right)$ matrix $\left(\underline{\underline{\chi}}^{ECT-UT}\right)'$ is built by subtracting to each f-th ($f = 1, ..., F^{ECT-UT}$) column of $\underline{\underline{\chi}}^{ECT-UT}$ its mean value μ_f. Similarly, a $(N_0 \times Q)$ matrix $\underline{\underline{p}}'$ is built by subtracting to each

q-th column of the parameter matrix $\underline{\underline{p}}$ its mean value μ_q. Apply the PLS algorithm to linearly decompose $\left(\underline{\underline{\chi}}^{ECT-UT}\right)'$ and $\underline{\underline{p}}'$ as follows

$$\left(\underline{\underline{\chi}}^{ECT-UT}\right)' = \underline{\underline{T}}^{ECT-UT} \times \underline{\underline{S}} + \underline{\underline{Y}}; \quad \underline{\underline{p}}' = \underline{\underline{U}} \times \underline{\underline{Z}} + \underline{\underline{G}}. \tag{3}$$

$\underline{\underline{T}}^{ECT-UT} = \left\{\left(\underline{T}^{ECT-UT}\right)^{(n)}; n = 1, ..., N_0\right\}$ in Eq. (3) is the $(N \times J)$ matrix of χ-scores $[\left(\underline{T}^{ECT-UT}\right)^{(n)} = \left\{\left(T_j^{ECT-UT}\right)^{(n)}; j = 1, ..., J\right\}]$. It is obtained from $\left(\underline{\underline{\chi}}^{ECT-UT}\right)'$ through the $\left(F^{ECT-UT} \times J\right)$ weight matrix $\underline{\underline{W}}$ [i.e., $\underline{\underline{T}}^{ECT-UT} = \left(\underline{\underline{\chi}}^{ECT-UT}\right)' \times \underline{\underline{W}}$]. Among different iterative algorithms, we have used SIMPLS algorithm [17] to obtain the weight matrix $\underline{\underline{W}}$. $\underline{\underline{Y}}$ and $\underline{\underline{G}}$ contain the $\left(N_0 \times F^{ECT-UT}\right)$ and $(N_0 \times Q)$ residuals of the linear decomposition while $\underline{\underline{S}}$ and $\underline{\underline{Z}}$ are the $\left(J \times F^{ECT-UT}\right)$ and $(J \times Q)$ matrices of loadings. The decomposition in Eq. (3) is aimed at maximizing the co-variance between the corresponding columns of $\underline{\underline{T}}^{ECT-UT}$ and of the $(N_0 \times J)$ matrix of p-scores $\underline{\underline{U}}$. This guarantees all the information about the ECT-UT data embedded inside $\underline{\underline{\chi}}'$ (i.e., inside $\underline{\underline{\chi}}$) is compressed into $\underline{\underline{T}}$. Assign the number of training sample $N_{iterative} = N_0$ and construct an initial training set $\hat{D}_{N_{iterative}} = \left[\left\{\left(\underline{T}^{ECT-UT}\right)^{(n)}; \underline{p}^{(n)}\right\}; n = 1, ..., N_{iterative}\right]$ for the adaptive step.

iii **Adaptive Sampling-** Generate V candidate samples by $\underline{p}_{cand}^{(v)} = \left(p_{cand,q}^{(v)}; q = 1, ..., Q\right)$ through Latin Hypercube Sampling (LHS) strategy where $v = 1, ..., V$. An estimation of the J-dimensional set of extracted features corresponding to each v-th candidate, $\left(\underline{\tilde{T}}_{cand}^{ECT-UT}\right)^{(v)}$ is retrieved by applying a multi-dimensional linear interpolator on $\hat{D}_{N_{iterative}}$. Select the optimal $v = v_{opt}$ candidate (i.e., $\underline{p}_{cand}^{(v_{opt})}$) from V such that the minimum distance between the obtained extracted features $\left(\underline{\tilde{T}}_{cand}^{ECT-UT}\right)^{(v_{opt})}$ and all the available extracted features $\left(\underline{T}^{ECT-UT}\right)^{(n)}$ $(n = 1, ..., N_{iterative})$ within $\hat{D}_{N_{iterative}}$ is maximized [i.e., $v_{opt} = \arg\left(\max_{v=1,...,V}\left\{\min_{n=1,...,N}\left[d_{vn}\right]\right\}\right)$]. d_{vn} is the Euclidean distance between $\left(\underline{\tilde{T}}_{cand}^{ECT-UT}\right)^{(v)}$ and $\left(\underline{T}^{ECT-UT}\right)^{(n)}$, which can be described by $d_{vn} = \sqrt{\sum_{j=1}^{J}\left\{\left(\underline{\tilde{T}}_{cand,j}^{ECT-UT}\right)^{(v)} - \left(\underline{T}_j^{ECT-UT}\right)^{(n)}\right\}^2}$. $\left(\underline{\chi}_{cand}^{ECT}\right)^{(v_{opt})}$ and $\left(\underline{\chi}_{cand}^{UT}\right)^{(v_{opt})}$ associated to the selected candidate sample are computed by utilizing $\Phi^{ECT}\{.\}$ and $\Phi^{UT}\{.\}$, respectively and obtained the corresponding ECT-UT features $\left(\underline{\chi}_{cand}^{ECT-UT}\right)^{(v_{opt})}$. The set of extracted features is obtained by $\left(\underline{T}_{cand}^{ECT-UT}\right)^{(v_{opt})} = \left\{\left(\underline{\chi}_{cand}^{ECT-UT}\right)^{(v_{opt})}\right\}' \times \underline{\underline{W}}$. Update the training set $\hat{D}_{N_{iterative}+1} = \hat{D}_{N_{iterative}} \cup \left\{\left(\underline{T}_{cand}^{ECT-UT}\right)^{(v_{opt})}; \underline{p}_{cand}^{(v_{opt})}\right\}$ with $N_{iterative} = N_{iterative} + 1$.

iv **Stop Criterion-** The adaptive sampling step adds new sample iteratively until $N_{iterative} = N$ (N is desired/feasible training size).

At this stage, an ε-SVR [12] has been utilized to train separately q-th set of I/O pairs $\hat{D}_{N,q} = \left[\left\{\left(\underline{T}^{ECT-UT}\right)^{(n)}; p_q^{(n)}\right\}; n = 1, ..., N\right]$ on the generated training set for each q-th parameter $(q = 1, ..., Q)$ of the crack. The m-th test sample $\left(\underline{\chi}^{ECT-UT}\right)^{(m)}$ of F^{ECT-UT} ECT-UT features associated to a previously-unseen crack parameter configuration $\underline{p}^{(m)}$ is projected through $\underline{\underline{W}}$ into the J-dimensional PLS-extracted features space [i.e., $\left(\underline{T}^{ECT-UT}\right)^{(m)} = \left\{\left(\underline{\chi}^{ECT-UT}\right)^{(m)}\right\}' \times \underline{\underline{W}}$]. Finally, the q-th crack parameter associated to $\left(\underline{T}^{ECT-UT}\right)^{(m)}$ is estimated (i.e., $\tilde{p}_q^{(m)}$) by the corresponding trained model in online phase.

3. Numerical validation

The ECT probe and the UT probe collect their corresponding NDT data from 81 positions along X directions with a step size of 0.5 mm and from 41 positions along Y directions with a step size of 1 mm, respectively through a raster scan. Therefore, ECT (i.e., impedance variation signal) and UT signals (i.e., reflected rays) are collected from $K = 81 \times 41 = 3321$ number of inspected points. Therefore, for a single crack configuration (i.e., sample), $F^{ECT} = 2K = 6642$, $F^{UT} = K = 3321$ and $F^{ECT-UT} = 9963$ actual features are treated. For having a valid comparison, $J = 20$ most significant features are extracted from each of these higher dimensional data sets by PLS feature extraction strategy. Different training sets (corresponding to three different crack parameters) have been created by changing the crack dimensions within the range $l_c \in [3.00, 10.00]$ mm, $\delta_c \in [0, 4.00]$ mm and $\varphi_c \in [0, 90]$ deg. by PLS-OSF sampling approach through utilizing ECT, UT and ECT-UT data separately. The initial and maximum number of samples are chosen for $N_0 = 27$ and $N = 216$, respectively. Figure 2 represents the

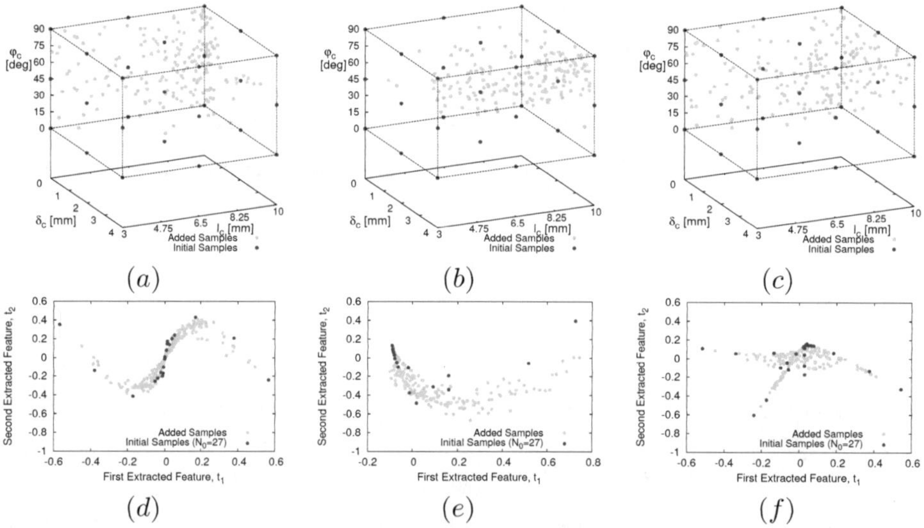

Figure 2. Training samples mapped on (a)-(c) the crack parameter space, (d)-(f) the extracted feature space for (a)-(d) ECT, (b)-(e) UT and (c)-(f) ECT-UT while $N_0 = 27$, $J = 2$, $N = 216$.

distribution of the resultant training samples in the parameter space as well as in the extracted feature space (for imaging purpose, the first 2 extracted features are considered) of all the data sets (i.e., ECT, UT and ECT-UT). $M = 1000$ unknown samples for 3 crack parameters have been generated by using LHS design and the corresponding 3 test sets of ECT, UT and ECT-UT actual features are obtained. $J = 20$ features are extracted by projecting the test sets into extracted feature space through the PLS weight matrix (e.g., obtained from the corresponding ECT, UT and ECT-UT methods during training phase) for each test set. To partially consider noise effects, Additive White Gaussian Noise (AWGN) has been imposed for different signal to noise ratio (SNR) (e.g., $[10, 20, 30, 40]$ [dB]) for blurring ECT, UT signals separately, obtain the corrupted ECT-UT features and project to the extracted feature space. Normalized mean error

(NME) described in [4] has been utilized for evaluating the inversion performance.

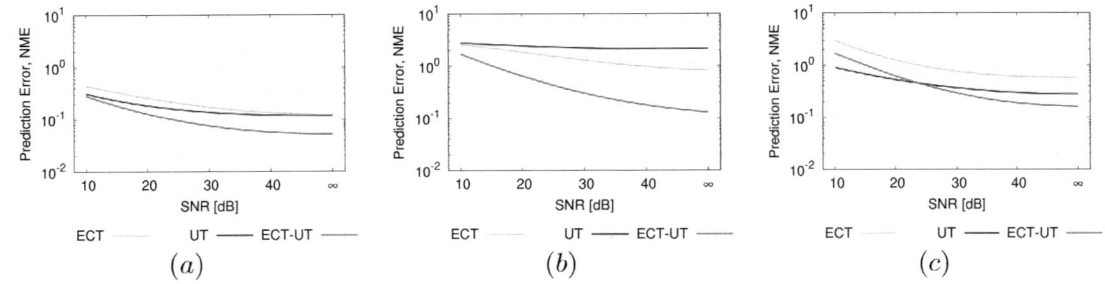

Figure 3. NME vs. SNR representation for crack (a) length l_c, (b) ligament δ_c and (c) angular position φ_c estimation for $N = 216$, $J = 20$, $M = 1000$ through ECT, UT and ECT-UT data.

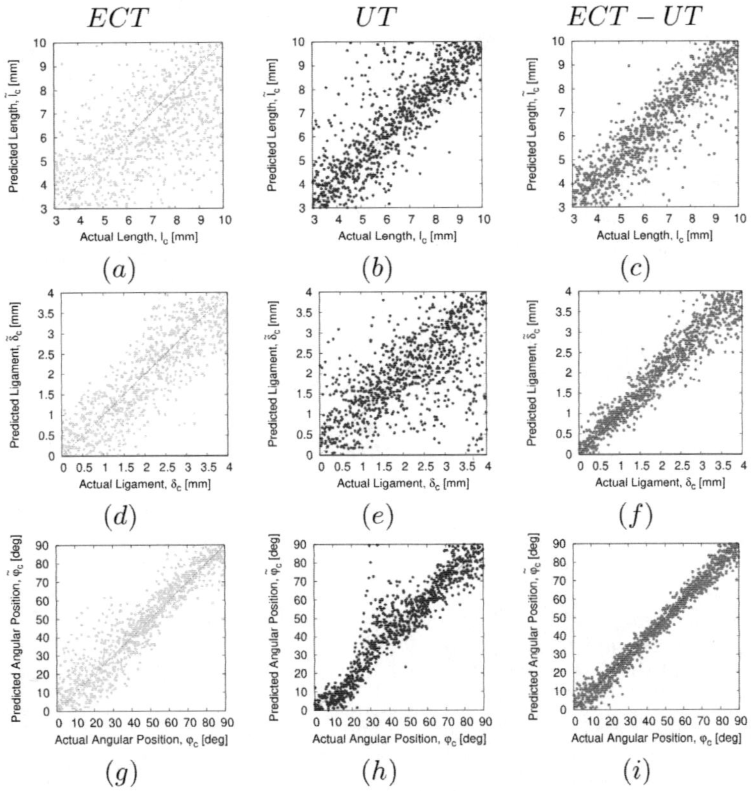

Figure 4. Actual vs. predicted plots for M=1000 test configurations at $SNR = 20$ [dB], for $N = 216$, $J = 20$, $M = 1000$ for crack (a)-(c) length l_c, (d)-(f) ligament δ_c and (g)-(i) angular position φ_c through ECT, UT and ECT-UT.

ECT signals are mostly corrupted for imposing noise and by combining ECT and UT signals, we can improve the overall inversion performance. Fig. 3 (a) shows that l_c estimation by using ECT suffers on noisy data than UT, while on $Noiseless$ test set by both ECT and UT have shown similar prediction accuracy. On the other hand, crack ligament distance δ_c estimation is showing lower prediction error for adopting ECT signals than UT signals for both noisy and $Noiseless$ test set (Fig. 3 (b)). Whereas, UT data shows lower NME than ECT data for

IOP Conf. Series: Journal of Physics: Conf. Series **1131** (2018) 012012 doi:10.1088/1742-6596/1131/1/012012

angular distance φ_c estimation (Fig. 3 (*c*)). By combining both ECT and UT signals, ECT-UT data fusion contains both information from ECT and UT signals. Whereas, applying PLS-OSF sampling, we can retrieve most significant information from ECT-UT. As a consequence, it improves the learning ability of SVR during training model development. Hence, ECT-UT data fusion has shown higher prediction accuracy than ECT and UT data for all the cases on noisy and $Noiseless$ test sets. Fig. 4 shows the scatter plots of true vs. predicted crack parameters obtained for $N = 216$ for noisy test set ($SNR = 20$ [dB]). Qualitatively, ECT-UT data fusion provides better l_c, δ_c and φ_c estimation than ECT and UT signals. Concerning real time solution, it takes 0.03s for testing 1000 samples during online phase.

4. Conclusions

In this work, we have shown an innovative MP-LBE inversion strategy for crack dimension and position estimation. Within the framework of LBE, PLS-OSF/SVR strategy has been applied for solving a NDE problem by utilizing ECT signals, UT signals and ECT-UT data fusion. By combining two different NDT methods, we first retrieved the variation of actual ECT and UT signals for changing crack parameters. Applying adaptive sampling through PLS feature extraction retrieves most significant information from the actual ECT-UT features that improves the learning ability of SVR. ECT-UT shows better prediction accuracy than ECT and UT methods separately for performing inversion on both noisy and noiseless synthetic test set.

References

[1] Salucci M, Poli L, Anselmi N and Massa A 2017 *IEEE Transactions on Geoscience and Remote Sensing* **55** 1305–1317
[2] Salucci M, Poli L and Massa A 2017 *Signal Processing* **132** 306 – 318 ISSN 0165-1684
[3] Salucci M, Oliveri G and Massa A 2015 *IEEE Transactions on Geoscience and Remote Sensing* **53** 6573–6592 ISSN 0196-2892
[4] Ahmed S, Salucci M, Miorelli R, Anselmi N, Oliveri G, Calmon P, Reboud C and Massa A 2017 *Journal of Physics: Conference Series* **904**
[5] Ahmed S, Miorelli R, Salucci M and Massa A 2017 *Studies in Applied Electromagnetics and Mechanics: Electromagnetic Nondestructive Evaluation (XX)* **42** 228–235
[6] Salucci M, Vrba J, Merunka I and Massa A 2017 *Microwave and Optical Technology Letters* **59** 2796–2799 ISSN 1098-2760
[7] Massa A, Oliveri G, Salucci M, Anselmi N and Rocca P 2018 *Journal of Electromagnetic Waves and Applications* **32** 516–541
[8] Pipis K, Skarlatos A, Theodoulidis T and Lesselier D 2016 *IEEE Transactions on Magnetics* **52** 1–8 ISSN 0018-9464
[9] Liu Z, Forsyth D S, Komorowski J P, Hanasaki K and Kirubarajan T 2007 *IEEE Transactions on Instrumentation and Measurement* **56** 2435–2451
[10] Wold S, Sjostrom M and Eriksson L 2001 *Chemometrics and Intelligent Laboratory Systems* **58** 109–130
[11] Bilicz S, Lambert M and Gyimothy S 2010 *Inverse Problems* **26** 074012
[12] Smola A J and Schölkopf B 2004 *Statistics and Computing* **14** 199–222 ISSN 0960-3174
[13] Lesselier D and Razek A 2002 *Scattering: Scattering and Inverse Scattering in Pure and Applied Science: Part 1 – Scattering of Waves by Macroscopic Targets,* (Academic Press) pp 486–507
[14] CEA-LIST CIVA: Simulation and Analysis for NDT http://www-civa.cea.fr/en/ [Online; accessed Mar. 2018]
[15] Hopkins D, Datuin M, Aldrin J, Warchol M, Warchol L and Forsyth D 2017 *AIP Conference Proceedings* **1806** 090007
[16] Zernov V, Fradkin L and Darmon M 2012 *Ultrasonics* **52** 830–835
[17] De Jong S 1993 *Chemometrics and Intelligent Laboratory Systems* **18** 251–263

Acknowledgments

This work has been partially supported by the SIRENA project (2014-2017) funded by DIGITEO (France) under the Call for Chairs 2014, the French project ANR-ByPASS, and by the Italian Ministry of Foreign Affairs and International Cooperation, Directorate General for Cultural and Economic Promotion and Innovation within the SNATCH Project (2017-2019).

IOP Conf. Series: Journal of Physics: Conf. Series **1131** (2018) 012013 doi:10.1088/1742-6596/1131/1/012013

A Bregman inexact linesearch–based forward–backward algorithm for nonsmooth nonconvex optimization

Simone Rebegoldi[1], **Silvia Bonettini**[2] **and Marco Prato**[2]

[1] Dipartimento di Matematica e Informatica, Università di Ferrara, Via Saragat 1, 44122 Ferrara, Italy

[2] Dipartimento di Scienze Fisiche, Informatiche e Matematiche, Università di Modena e Reggio Emilia, Via Campi 213/b, 41125 Modena, Italy

E-mail: simone.rebegoldi@unife.it

Abstract. In this paper, we present a forward–backward linesearch–based algorithm suited for the minimization of the sum of a smooth (possibly nonconvex) function and a convex (possibly nonsmooth) term. Such algorithm first computes inexactly the proximal operator with respect to a given Bregman distance, and then ensures a sufficient decrease condition by performing a linesearch along the descent direction. The proposed approach can be seen as an instance of the more general class of descent methods presented in [7], however, unlike in [7], we do not assume the strong convexity of the Bregman distance used in the proximal evaluation. We prove that each limit point of the iterates sequence is stationary, we show how to compute an approximate proximal–gradient point with respect to a Bregman distance and, finally, we report the good numerical performance of the algorithm on a large scale image restoration problem.

1. Introduction

This paper is concerned with the following optimization problem

$$\arg\min_{x\in\mathbb{R}^n} f(x) \equiv f_0(x) + f_1(x) \tag{1}$$

where $f_0 : \mathbb{R}^n \to \mathbb{R}$ is continuously differentiable on an open set Ω, $f_1 : \mathbb{R}^n \to \mathbb{R}\cup\{\infty\}$ is a proper, convex, continuous function on a closed domain $\operatorname{dom} f_1 \subseteq \Omega$, and f is bounded from below. Several problems arising in signal and image processing can be reformulated as (1), such as image deconvolution, image segmentation, image compression and inpainting, non-negative matrix and tensor factorization [4, 10]. The structure of the objective function in (1) can be exploited by using the forward–backward (or proximal–gradient) algorithm [7, 12, 13], whose general iteration is given by

$$x^{(k+1)} = x^{(k)} + \lambda_k \left(\operatorname{prox}^{D_k}_{\alpha_k f_1}(x^{(k)} - \alpha_k D_k^{-1}\nabla f_0(x^{(k)})) - x^{(k)}\right), \tag{2}$$

where $\alpha_k > 0$ is a steplength parameter, $D_k \in \mathbb{R}^{n\times n}$ is a symmetric positive definite matrix, $\operatorname{prox}^{D}_{\alpha f_1}$ denotes the proximal operator of αf_1 w.r.t. the norm $\|x\|_D = x^T Dx$ induced by D, i.e.

$$\operatorname{prox}^{D}_{\alpha f_1}(z) = \arg\min_{y\in\mathbb{R}^n} f_1(y) + \frac{1}{2\alpha}\|y - z\|^2_D, \tag{3}$$

IOP Conf. Series: Journal of Physics: Conf. Series **1131** (2018) 012013 doi:10.1088/1742-6596/1131/1/012013

and $\lambda_k \in (0,1)$ is an additional relaxation parameter which aims at ensuring the sufficient decrease of the objective function. Since, in its basic implementation, method (2) can be quite slow, several accelerated versions have been devised, in which either the steplength α_k and/or the scaling matrix D_k are chosen in a variable manner [7, 8, 11]. More recently, the scheme (2) has been generalized by replacing the (scaled) Euclidean norm in (3) with the more general concept of Bregman distance [1, 7, 9]. Given $\varphi : \mathbb{R}^n \to \mathbb{R}\cup\{\infty\}$ a strictly convex and differentiable function on $\mathrm{int}(\mathrm{dom}\,\varphi) \neq \emptyset$, the Bregman distance $D_\varphi : \mathrm{dom}(\varphi) \times \mathrm{int}(\mathrm{dom}\,\varphi) \to [0,+\infty]$ associated to φ is defined as [2, Definition 3.1]

$$D_\varphi(x,y) = \varphi(x) - \varphi(y) - \nabla\varphi(y)^T(x-y), \quad \forall\, x \in \mathrm{dom}(\varphi), \quad \forall\, y \in \mathrm{int}(\mathrm{dom}\,\varphi). \tag{4}$$

Observe that the scaled Euclidean norm is recovered from (4) by setting $\varphi = \|\cdot\|_D^2/2$. Then one can apply (2) by using, instead of (3), the following generalized proximal operator [1, p. 337]

$$\mathrm{prox}_{\alpha f_1}^{\varphi}(z) = \arg\min_{y\in\mathbb{R}^n} f_1(y) + \frac{1}{\alpha} D_\varphi(y,z). \tag{5}$$

In the following, we present a Bregman version of algorithm (2) whose main features are the computation of an approximate proximal–gradient point $\tilde{y}^{(k)}$ with respect to a prefixed Bregman distance, the adaptive choice of the steplength α_k, which can take any value in a prefixed interval $[\alpha_{\min}, \alpha_{\max}]$, and the linesearch performed along the descent direction detected by the approximate point $\tilde{y}^{(k)}$. Such algorithm can be considered as an instance of the general framework proposed in [7], where the concept of Bregman distance is replaced by a generalized distance-like function satisfying appropriate assumptions. We prove that each limit point of the iterates sequence is stationary under the assumption of super-coercivity of the Bregman distance, which is weaker than the one of strong convexity adopted in [7]. Furthermore, we provide a general procedure to compute an approximate Bregman proximal–gradient point satisfying the required inexactness criterion, and we evaluate the effectiveness of the proposed approach on a Poisson image restoration problem.

2. Algorithm and convergence result

The domain of a function $f : \mathbb{R}^n \to \mathbb{R}\cup\{\infty\}$ is defined as $\mathrm{dom} f = \{x \in \mathbb{R}^n : f(x) < \infty\}$. Given $\epsilon > 0$, the ϵ−subdifferential of a proper, convex function f at $x \in \mathrm{dom} f$ is the set $\partial_\epsilon f(x) = \{w \in \mathbb{R}^n : f(z) \geq f(x) + w^T(z-x) - \epsilon,\ \forall\, z \in \mathbb{R}^n\}$. The limiting subdifferential of a function f at $x \in \mathrm{dom} f$ is defined as $\partial f(x) = \{v \in \mathbb{R}^n : \exists \{y^{(k)}\}_{k\in\mathbb{N}} \subseteq \mathbb{R}^n, v^{(k)} \in \hat{\partial} f(y^{(k)})\ \forall k \in \mathbb{N} \text{ s.t. } y^{(k)} \to x,\ f(y^{(k)}) \to f(x) \text{ and } v^{(k)} \to v\}$, where $\hat{\partial} f(x)$ denotes the Fréchet subdifferential of f at x [8, Definition 1]. The conjugate function f^* is defined as $f^*(v) = \sup_x\, x^T v - f(x)$. For any $c \in \mathbb{R}$, $\mathrm{lev}_{\leq c}(f) = \{x \in \mathbb{R}^n : f(x) \leq c\}$ denotes the sublevel set of f.
Throughout the paper, we will make the following blanket assumptions.

Assumption 1 *(i) φ is a Legendre function [2], i.e. it is strictly convex and differentiable on* $\mathrm{int}(\mathrm{dom}\,\varphi)$*, and* $\|\nabla\varphi(x^{(k)})\| \to \infty$ *for every sequence* $\{x^{(k)}\} \subseteq \mathrm{int}(\mathrm{dom}\,\varphi)$ *converging to a boundary point of* $\mathrm{dom}\,\varphi$ *as* $k \to \infty$.

(ii) $0 \in \mathrm{int}(\mathrm{dom}\,\varphi)$.

(iii) $\mathrm{dom}\, f_1 \subseteq \mathrm{int}(\mathrm{dom}\,\varphi)$.

(iv) $\mathrm{dom}\,\varphi^* = \mathbb{R}^n$.

Remark 1 *Note that Assumption 1(iv) is equivalent to require that φ is super-coercive, namely [2, Definition 2.15, Proposition 2.16]*

$$\lim_{\|x\|\to+\infty} \frac{\varphi(x)}{\|x\|} = +\infty. \tag{6}$$

Such an hypothesis is weaker than requiring strong convexity, i.e. that $\varphi - \frac{m}{2}\|\cdot\|^2$ is convex for a certain $m > 0$. Indeed every strongly convex function is super-coercive [3, Corollary 11.17] but the converse is not true. As an example, consider the Boltzmann–Shannon entropy $\varphi(x) = \sum_{i=1}^n \varphi_i(x_i) = \sum_{i=1}^n (x_i + \gamma)\log(x_i + \gamma)$, where $\gamma > 0$ and $0\log 0 = 0$ [1]. Note that $\varphi^*(v) = \sum_{i=1}^n (e^{v_i - 1} - \gamma v_i)$ is defined on all $\mathbb{R}^n$ and, therefore, φ is super-coercive. However, the second derivative of φ_i is bounded away from zero only on bounded subsets of the half-line $(-\gamma, +\infty)$, hence φ is not strongly convex.

We also recall the following properties of Bregman distances.

Theorem 1 *[2] Suppose that $\varphi : \mathbb{R}^n \to \mathbb{R} \cup \{\infty\}$ satisfies Assumption 1. Then*

(i) $\nabla\varphi : \mathrm{int}(\mathrm{dom}\,\varphi) \to \mathrm{int}(\mathrm{dom}\,\varphi^*)$ is a bijection and $(\nabla\varphi)^{-1} = \nabla\varphi^*$.

(ii) For all $x, y \in \mathrm{int}(\mathrm{dom}\,\varphi)$, we have

$$D_\varphi(x, y) = D_{\varphi^*}(\nabla\varphi(y), \nabla\varphi(x)). \tag{7}$$

(iii) (Three Points Identity) For all $x \in \mathrm{dom}(\varphi)$, $y, z \in \mathrm{int}(\mathrm{dom}\,\varphi)$, we have

$$D_\varphi(x, z) = D_\varphi(x, y) + D_\varphi(y, z) + (\nabla\varphi(y) - \nabla\varphi(z))^T(x - y). \tag{8}$$

Given $x^{(k)} \in \mathrm{dom}\, f_1$ and $\alpha_k > 0$, we define the following function

$$h_{\alpha_k}(y, x^{(k)}) := \nabla f_0(x^{(k)})^T(y - x^{(k)}) + \frac{1}{\alpha_k} D_\varphi(y, x^{(k)}) + f_1(y) - f_1(x^{(k)}) \tag{9}$$

and, assuming that (9) has an unique minimizer, we set $y^{(k)} = \arg\min_{y\in\mathbb{R}^n} h_{\alpha_k}(y, x^{(k)})$. As described in [1, Section 3.1], the point $y^{(k)}$ can be naturally decomposed as a Bregman proximal–gradient point. Indeed, let us write down the optimality condition

$$0 \in \partial h_{\alpha_k}(y^{(k)}, x^{(k)}) \Leftrightarrow 0 \in \alpha_k \nabla f_0(x^{(k)}) + \nabla\varphi(y^{(k)}) - \nabla\varphi(x^{(k)}) + \alpha_k \partial f_1(y^{(k)}). \tag{10}$$

Since Assumption 1(iv) implies that $\nabla\varphi(x^{(k)}) - \alpha_k \nabla f_0(x^{(k)}) \in \mathrm{dom}\,\nabla\varphi^*$, we can always define the point $z^{(k)} = \nabla\varphi^*(\nabla\varphi(x^{(k)}) - \alpha_k \nabla f_0(x^{(k)}))$. Then, by recalling Theorem 1(i), the optimality condition (10) can be equivalently rewritten as

$$0 \in \nabla\varphi(y^{(k)}) - \nabla\varphi(z^{(k)}) + \alpha_k \partial f_1(y^{(k)})$$

which yields $y^{(k)} = \mathrm{prox}^{\varphi}_{\alpha_k f_1}(z^{(k)})$.

We are now ready to fully detail the proposed approach, which is reported in Algorithm 1. At each iteration, we look for an approximation $\tilde{y}^{(k)}$ of the Bregman proximal–gradient point $y^{(k)}$ which satisfies the following conditions

$$h_{\alpha_k}(\tilde{y}^{(k)}, x^{(k)}) < 0, \quad 0 \in \partial_{\epsilon_k} h_{\alpha_k}(\tilde{y}^{(k)}, x^{(k)}). \tag{11}$$

On one hand, the first condition guarantees that the point $\tilde{y}^{(k)}$ defines a descent direction $d^{(k)} = \tilde{y}^{(k)} - x^{(k)}$ [7, Proposition 2.2.] and, on the other hand, it ensures that the linesearch at STEP 3 is well-defined [7, Proposition 3.1]. The second condition can be seen as a relaxation of the optimality condition (10) associated to the point $y^{(k)}$, where the exact subdifferential of h_{α_k} is replaced with its ϵ_k–subdifferential. The linesearch at STEP 3 enforces condition (12), thus making $\{f(x^{(k)})\}_{k\in\mathbb{N}}$ a monotone nondecreasing sequence.

Algorithm 1 belongs to the class of line–search based descent methods described in [7], denominated Variable Metric Inexact Linesearch based Algorithms (VMILA). In [7, Theorem

Algorithm 1 Bregman Inexact Linesearch based Algorithm

Choose $x^{(0)} \in \operatorname{dom} f_1$, $0 \le \alpha_{\min} \le \alpha_{\max}$, $\varphi : \mathbb{R}^n \to \mathbb{R}\cup\{\infty\}$ satisfying Assumption 1, $\delta, \beta \in (0,1)$.
FOR $k = 0, 1, 2, ...$

STEP 1. Choose $\alpha_k \in [\alpha_{\min}, \alpha_{\max}]$ and $\epsilon_k > 0$.
STEP 2. Compute $\tilde{y}^{(k)}$ satisfying (11) and set $d^{(k)} = \tilde{y}^{(k)} - x^{(k)}$.
STEP 3. Compute the smallest non-negative integer i_k such that

$$f(x^{(k)} + \delta^{i_k} d^{(k)}) \le f(x^{(k)}) + \beta\delta^{i_k} h_{\alpha_k}(\tilde{y}^{(k)}, x^{(k)}) \tag{12}$$

and set $\lambda_k = \delta^{i_k}$.
STEP 4. Set $x^{(k+1)} = x^{(k)} + \lambda_k d^{(k)}$.

END

3.1], the stationarity of the limit points of VMILA is proved according to the following outline: (i) show that the subsequence $\{\tilde{y}^{(k)}\}_{k\in K}$ is bounded (using the strong convexity of φ); (ii) deduce that $\lim_{k\to\infty, k\in K} h_{\alpha_k}(\tilde{y}^{(k)}, x^{(k)}) = 0$ (using [7, Proposition 3.1]) ; (iii) conclude that the limit point is stationary (using again the strong convexity of φ). In the following theorem, we prove an analogous result for Algorithm 1, by adopting the same line of proof described above, but assuming that φ is super-coercive instead of strongly convex.

Theorem 2 *Assume that there exists a limit point $\bar{x}$ of $\{x^{(k)}\}_{k\in\mathbb{N}}$ and that $\lim_{k\to\infty} \epsilon_k = 0$. Then $\bar{x}$ is a stationary point for problem* (1)*, i.e.* $0 \in \partial f(\bar{x})$.

Proof: Let $K \subseteq \mathbb{N}$ be a subset of indices such that $\lim_{k\to\infty, k\in K} x^{(k)} = \bar{x}$ and $\lim_{k\to\infty,k\in K} \alpha_k = \bar{\alpha} \in [\alpha_{\min}, \alpha_{\max}]$. By continuity of the function $h_\alpha(y, x)$ with respect to the arguments α and x, we also have $\lim_{k\to\infty,k\in K} y^{(k)} = \bar{y}$, with $\bar{y} = \arg\min_y h_{\bar{\alpha}}(y, \bar{x})$. From (10), it follows $w \in \partial f_1(y^{(k)})$ if and only if $w = -\nabla f_0(x^{(k)}) - \frac{1}{\alpha_k}(\nabla\varphi(y^{(k)}) - \nabla\varphi(x^{(k)}))$. Observe that the right-hand side condition in (11) can be rewritten as $h_{\alpha_k}(\tilde{y}^{(k)}) - h_{\alpha_k}(y^{(k)}) \le \epsilon_k$. Then we have

$$\begin{aligned}
\epsilon_k &\ge h_{\alpha_k}(\tilde{y}^{(k)}) - h_{\alpha_k}(y^{(k)}) \qquad (13)\\
&= \nabla f_0(x^{(k)})^T(\tilde{y}^{(k)} - y^{(k)}) + \frac{1}{\alpha_k}\left(D_\varphi(\tilde{y}^{(k)}, x^{(k)}) - D_\varphi(y^{(k)}, x^{(k)})\right) + f_1(\tilde{y}^{(k)}) - f_1(y^{(k)})\\
&\ge \nabla f_0(x^{(k)})^T(\tilde{y}^{(k)} - y^{(k)}) + \frac{1}{\alpha_k}\left(D_\varphi(\tilde{y}^{(k)}, x^{(k)}) - D_\varphi(y^{(k)}, x^{(k)})\right) + w^T(\tilde{y}^{(k)} - y^{(k)})\\
&= \frac{1}{\alpha_k}\left(D_\varphi(\tilde{y}^{(k)}, x^{(k)}) - D_\varphi(y^{(k)}, x^{(k)}) - (\nabla\varphi(y^{(k)}) - \nabla\varphi(x^{(k)}))^T(\tilde{y}^{(k)} - y^{(k)})\right)\\
&\overset{(8)}{=} \frac{1}{\alpha_k} D_\varphi(\tilde{y}^{(k)}, y^{(k)}) \ge \frac{1}{\alpha_{\max}} D_\varphi(\tilde{y}^{(k)}, y^{(k)}).
\end{aligned}$$

Thanks to the hypothesis on $\{\epsilon_k\}_{k\in\mathbb{N}}$, the above inequalities yield $\lim_{k\to\infty} D_\varphi(\tilde{y}^{(k)}, y^{(k)}) = 0$. By applying (8) with $x = \tilde{y}^{(k)}$, $y = \bar{y}$, $z = y^{(k)}$, and the Cauchy-Schwarz inequality, we have

$$\begin{aligned}
D_\varphi(\tilde{y}^{(k)}, y^{(k)}) &= D_\varphi(\tilde{y}^{(k)}, \bar{y}) + D_\varphi(\bar{y}, y^{(k)}) + (\nabla\varphi(\bar{y}) - \nabla\varphi(y^{(k)}))^T(\tilde{y}^{(k)} - \bar{y})\\
&\ge D_\varphi(\tilde{y}^{(k)}, \bar{y}) - \|\nabla\varphi(\bar{y}) - \nabla\varphi(y^{(k)})\|\|\tilde{y}^{(k)} - \bar{y}\|
\end{aligned}$$

Since $\{y^{(k)}\}_{k\in K}$ is bounded, $\nabla\varphi$ is continuous and $\lim_{k\to\infty,k\in K} D_\varphi(\tilde{y}^{(k)}, \bar{y}) = 0$, there exist two constants $c_1, c_2 > 0$ such that

$$\frac{D_\varphi(\tilde{y}^{(k)}, \bar{y}) - c_1}{\|\tilde{y}^{(k)}\|} \le c_2, \quad \forall\, k \in K$$

IOP Conf. Series: Journal of Physics: Conf. Series **1131** (2018) 012013 doi:10.1088/1742-6596/1131/1/012013

or, in other words, $\{\tilde{y}^{(k)}\}_{k\in K} \subseteq \text{lev}_{\leq c_2}\left(\frac{D_\varphi(\cdot,\bar{y})-c_1}{\|\cdot\|}\right)$. By Assumption 1(iv) combined with Remark 1, this means that the sequence $\{\tilde{y}^{(k)}\}_{k\in K}$ lies in a sublevel set of a coercive function. Therefore, $\{\tilde{y}^{(k)}\}_{k\in K}$ is bounded and, consequently, we can apply [7, Proposition 3.1] to obtain that $\lim_{k\to\infty,k\in K} h_{\alpha_k}(\tilde{y}^{(k)}, x^{(k)}) = 0$. This last fact, combined with (13) and the assumption on $\{\epsilon_k\}_{k\in\mathbb{N}}$, implies $\lim_{k\to\infty,k\in K} h_{\alpha_k}(y^{(k)}, x^{(k)}) = h_{\bar{\alpha}}(\bar{y},\bar{x}) = 0$, which allows to conclude that $\bar{x}$ is stationary [7, Proposition 2.3]. □

3. Practical computation of the approximate point $\tilde{y}^{(k)}$

In this section, we show how we can practically compute a point $\tilde{y}^{(k)}$ satisfying (11) in the case when $f_1 = g\circ A$, being $A \in \mathbb{R}^{m\times n}$ and $g : \mathbb{R}^m \to \mathbb{R}\cup\{\infty\}$ a proper, convex function. The following procedure is a generalization of the one presented in [7, Section 4.2] for (scaled) Euclidean distances. Let us rewrite (10) in its equivalent primal-dual and dual formulations:

$$\min_{y\in\mathbb{R}^n} h_{\alpha_k}(y, x^{(k)}) = \min_{y\in\mathbb{R}^n}\max_{v\in\mathbb{R}^m} F_{\alpha_k}(y, v, x^{(k)}) = \max_{v\in\mathbb{R}^m} \Psi_{\alpha_k}(v, x^{(k)}).$$

By applying Theorem 1(i) to point $z^{(k)}$, we have $\nabla f_0(x^{(k)}) = \left(\nabla\varphi(x^{(k)}) - \nabla\varphi(z^{(k)})\right)/\alpha_k$. Using this relation and (8), we can rewrite the function $h_{\alpha_k}(\cdot, x^{(k)})$ as follows

$$h_{\alpha_k}(y, x^{(k)}) = \frac{1}{\alpha_k}\left(D_\varphi(y, z^{(k)}) - D_\varphi(x^{(k)}, z^{(k)})\right) + f_1(y) - f_1(x^{(k)}).$$

Recalling that $f_1(y) = \max_{v\in\mathbb{R}^m} v^T Ay - g^*(v)$, we come to the expression of $F_{\alpha_k}(y, v, x^{(k)})$:

$$F_{\alpha_k}(y, v, x^{(k)}) = \frac{1}{\alpha_k} D_\varphi(y, z^{(k)}) + v^T Ay - g^*(v) - \frac{1}{\alpha_k} D_\varphi(x^{(k)}, z^{(k)}) - f_1(x^{(k)}). \tag{14}$$

The minimum of $F_{\alpha_k}(\cdot, v, x^{(k)})$ is reached at the point $y^{(k)}_{\text{min}} = \nabla\varphi^*(\nabla\varphi(z^{(k)}) - \alpha A^T v)$. Replacing $y^{(k)}_{\text{min}}$ into $F_{\alpha_k}(y, v, x^{(k)})$, we get the dual function $\Psi_{\alpha_k}(\cdot, x^{(k)})$:

$$\begin{aligned}
\Psi_{\alpha_k}(v, x^{(k)}) &= \frac{1}{\alpha_k} D_\varphi(y^{(k)}_{\text{min}}, z^{(k)}) + v^T Ay^{(k)}_{\text{min}} - g^*(v) - \frac{1}{\alpha_k} D_\varphi(x^{(k)}, z^{(k)}) - f_1(x^{(k)})\\
&\overset{(8)}{=} \frac{1}{\alpha_k} D_\varphi(\nabla\varphi^*(\nabla\varphi(z^{(k)}) - \alpha_k A^T v), 0) + \frac{1}{\alpha_k}(\nabla\varphi(0) - (\nabla\varphi(z^{(k)}) - \alpha_k A^T v))^T(\nabla\varphi^*(\nabla\varphi(z^{(k)}) - \alpha_k A^T v))\\
&\quad - g^*(v) - \frac{1}{\alpha_k} D_\varphi(x^{(k)}, z^{(k)}) + \frac{1}{\alpha_k} D_\varphi(0, z^{(k)}) - f_1(x^{(k)})\\
&\overset{(7)}{=} \frac{1}{\alpha_k} D_{\varphi^*}(\nabla\varphi(0), \nabla\varphi(z^{(k)}) - \alpha_k A^T v)\\
&\quad + \frac{1}{\alpha_k}(\nabla\varphi(0) - (\nabla\varphi(z^{(k)}) - \alpha_k A^T v))^T(\nabla\varphi^*(\nabla\varphi(z^{(k)}) - \alpha_k A^T v) - \nabla\varphi^*(\nabla\varphi(0)))\\
&\quad - g^*(v) - \frac{1}{\alpha_k} D_\varphi(x^{(k)}, z^{(k)}) + \frac{1}{\alpha_k} D_\varphi(0, z^{(k)}) - f_1(x^{(k)})\\
&\overset{(8)}{=} -\frac{1}{\alpha_k} D_{\varphi^*}(\nabla\varphi(z^{(k)}) - \alpha_k A^T v, \nabla\varphi(0)) - g^*(v) - \frac{1}{\alpha_k} D_\varphi(x^{(k)}, z^{(k)}) + \frac{1}{\alpha_k} D_\varphi(0, z^{(k)}) - f_1(x^{(k)})
\end{aligned}$$

where, in the second equality, we used (8) with $x = y^{(k)}_{\text{min}}$, $y = 0$, $z = z^{(k)}$ whereas, in the fourth equality, we used (8) with $x = z = \nabla\varphi(0)$, $y = \nabla\varphi(z^{(k)}) - \alpha_k A^T v$. If it is possible to generate a sequence $\{v^{(k,\ell)}\}_{\ell\in\mathbb{N}}$ such that $v^{(k,\ell)} \underset{\ell\to\infty}{\longrightarrow} \arg\max_v \Psi_{\alpha_k}(v, x^{(k)})$,

$\Psi_{\alpha_k}(v^{(k,\ell)}, x^{(k)}) \underset{\ell\to\infty}{\longrightarrow} \max_v \Psi_{\alpha_k}(v, x^{(k)})$, then a point $\tilde{y}^{(k)}$ satisfying (11) can be found by setting $\tilde{y}^{(k,\ell)} = P_{\mathrm{dom}\, f_1}\left(\nabla\varphi^*(\nabla\varphi(z^{(k)}) - \alpha_k A^T v^{(k,\ell)})\right)$ and stopping the iterates when condition

$$h_{\alpha_k}(\tilde{y}^{(k,\ell)}, x^{(k)}) \le \eta \Psi_{\alpha_k}(v^{(k,\ell)}, x^{(k)}) \tag{15}$$

is satisfied, with $\eta \in (0, 1]$ being a prefixed parameter. Indeed, $\tilde{y}^{(k)} = \tilde{y}^{(k,\ell)}$ satisfies (11) with $\epsilon_k = (1 - 1/\eta) h_{\alpha_k}(\tilde{y}^{(k)}, x^{(k)})$ [7, Section 4.5].

4. Application to Poisson image restoration

In the Bayesian framework, the recovering of an unknown image $x \in \mathbb{R}^n$ from a blurred noisy image $g \in \mathbb{R}^n$ can be reformulated as a problem of the form (1), where f_0 is a discrepancy function, typically depending on the type of noise affecting the data, whereas f_1 is a regularization term including a-priori information and possible physical constraints. When the data are corrupted by Poisson noise, f_0 is the generalized Kullback-Leibler (KL) divergence [4]:

$$f_0(x) = KL(Hx + b; g) = \sum_{i=1}^{n} g_i \log \frac{g_i}{(Hx)_i + b} + (Hx)_i + b - g_i \tag{16}$$

where the matrix $H \in \mathbb{R}^{n\times n}$ represents the blurring operator satisfying the conditions $H_{i,j} \ge 0$, $H\mathbf{1}$ and $H^T\mathbf{1}$ (with $\mathbf{1}$ being the vector of all ones) and $b > 0$ is a background term. In order to preserve the edges of the image and the non negativity of its pixels, the term f_1 is chosen as

$$f_1(x) = \rho TV(x) + \iota_{\mathbb{R}^n_{\ge 0}}(x), \tag{17}$$

where $TV(x) = \rho \sum_{i=1}^n \|\nabla_i x\|_2$ is the total variation functional [14], being $\nabla_i \in \mathbb{R}^{2\times n}$ the discrete gradient operator at the i–th pixel, $\iota_{\mathbb{R}^n_{\ge 0}}$ is the indicator function of the non negative orthant and $\rho > 0$ is the regularization parameter.

We now discuss the application of Algorithm 1 to the above problem. Concerning the Legendre function φ, it seems quite natural, looking at the discrepancy term (16), to use the Boltzmann–Shannon entropy described in Remark 1. Indeed, the Bregman distance associated to such function is $D_\varphi(x, y) = \sum_{i=1}^n (x_i + \gamma) \log\left(\frac{x_i+\gamma}{y_i+\gamma}\right) + (y_i + \gamma) - (x_i + \gamma) = KL(y + \gamma\mathbf{1}; x + \gamma\mathbf{1})$, i.e. the Kullback–Leibler divergence itself. The approximate point $\tilde{y}^{(k)}$ is then computed as described in Section 3, using the scaled gradient projection method [6] to generate the dual sequence $\{v^{(k,\ell)}\}_{\ell\in\mathbb{N}}$. Regarding the steplength α_k, we observe that the point $z^{(k)}$ can be rewritten as $z_i^{(k)} = x_i^{(k)} - D_k(\alpha_k)\nabla_i f_0(x^{(k)}) = x_i^{(k)} - q_i(\alpha_k)(x_i + \gamma)\nabla_i f_0(x^{(k)})$, with $q_i(\alpha) = (1 - e^{-\alpha\nabla_i f_0(x^{(k)})})/\nabla_i f_0(x^{(k)})$ when $\nabla_i f_0(x^{(k)}) \ne 0$ and $q_i(\alpha) = \alpha$ when $\nabla_i f_0(x^{(k)}) = 0$. Thus, it is reasonable to consider the following quasi-Newton approaches

$$\alpha_1^{(k)} = \min_{\alpha\in[\alpha_{\min},\alpha_{\max}]} \|D_k(\alpha)^{-1} s^{(k)} - w^{(k)}\|^2, \quad \alpha_2^{(k)} = \min_{\alpha\in[\alpha_{\min},\alpha_{\max}]} \|s^{(k)} - D_k(\alpha) w^{(k)}\|^2$$

where $s^{(k)} = x^{(k)} - x^{(k-1)}$ and $w^{(k)} = \nabla f_0(x^{(k)}) - \nabla f_0(x^{(k-1)})$. In our experiments, we alternate the two above rules using the same strategy adopted in [6].

We show the results obtained by running Algorithm 1 on a set of three test problems well-known in the literature, whose description can be found in [7]. In Figure 1 we report the relative decrease of the objective function with respect to the iteration number of Algorithm 1, denominated VMILA-Bregman, in comparison with two other instances of the VMILA algorithm, one in which the proximal operator (5) is computed w.r.t to the Euclidean norm (VMILA-ID), and the other one in which a variable scaled Euclidean norm is used (VMILA-SG). As it can be drawn from the plots, both VMILA-Bregman and VMILA-SG outperform VMILA-ID of several orders of magnitude, which suggests that adopting non Euclidean metrics may be advantageous both in terms of efficiency and accuracy.

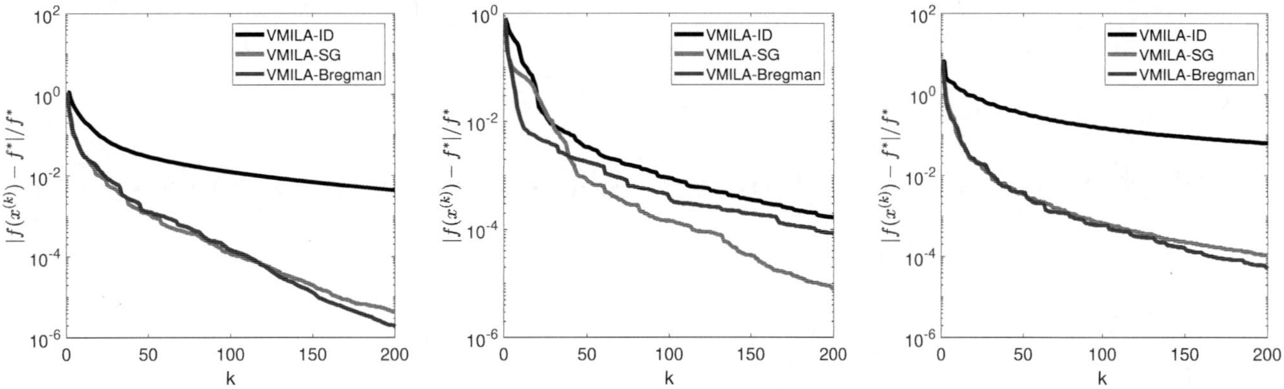

Figure 1. Relative decrease of the objective function values with respect to the iteration number. Left column: cameraman. Middle column. micro. Right column: phantom.

5. Acknowledgements
The authors are members of the INdAM Research group GNCS. This work has been partially supported by the Italian GNCS-INdAM under the project GNCS - Finanziamento Giovani Ricercatori 2017-2018.

References
[1] H. H. Bauschke, J. Bolte, and M. Teboulle 2017, A descent lemma beyond Lipschitz gradient continuity: first-order methods revisited and applications, *Math. Oper. Res.* **4**(1), 330–348.
[2] H. H. Bauschke, and J. M. Borwein 1997, Legendre functions and the method of Bregman projection, *J. Convex Anal.* **4**, 27–67.
[3] H. H. Bauschke, and P. L. Combettes 2011, *Convex analysis and monotone operator theory in Hilbert spaces*, (Springer).
[4] M. Bertero, and P. Boccacci 1998, *Introduction to Inverse Problems in Imaging*, (Bristol: Institute of Physics Publishing).
[5] M. Bertero, P. Boccacci, G. Desider, and G. Vicidomini 2009, Image deblurring with Poisson data: From cells to galaxies, *Inverse Probl.* **25**(12), 123006.
[6] S. Bonettini, R. Zanella, and L. Zanni 2009, A scaled gradient projection method for constrained image deblurring, *Inverse Probl.* **25**, 015002.
[7] S. Bonettini, I. Loris, F. Porta, and M. Prato 2016, Variable metric inexact line-search-based methods for nonsmooth optimization, *SIAM J. Optim.* **26**(2), 891–921.
[8] S. Bonettini, I. Loris, F. Porta, M. Prato, and S. Rebegoldi 2017, On the convergence of a linesearch based proximal–gradient method for nonconvex optimization, *Inverse Probl.* **33**(5), 055005.
[9] S. Bonettini, M. Prato, and S. Rebegoldi 2016, A cyclic block coordinate descent method with generalized gradient projections, *Appl. Math. Comput.* **286**, 288–300.
[10] A. Chambolle, and T. Pock 2016, An introduction to continuous optimization for imaging, *Acta Numerica* **25**, 161–319.
[11] E. Chouzenoux, J.-C. Pesquet, and A. Repetti 2014, Variable metric forward–backward algorithm for minimizing the sum of a differentiable function and a convex function, *J. Optim. Theory Appl.* **162**(1), 107–132.
[12] P. L. Combettes, and V. R. Wajs 2005, Signal recovery by proximal forward–backward splitting, *Multiscale Model. Simul.* **4**, 1168–1200.
[13] P. L. Combettes, and J.-C. Pesquet 2011, Proximal splitting methods in signal processing, in *Fixed-point algorithms for inverse problems in science and engineering*, H. H. Bauschke et al., New York, Springer, 185–212.
[14] L. Rudin, S. Osher, and E. Fatemi 1992, Nonlinear total variation based noise removal algorithms, *J. Phys. D.* **60**(1-4), 259–268.

IOP Conf. Series: Journal of Physics: Conf. Series **1131** (2018) 012014 doi:10.1088/1742-6596/1131/1/012014

Multiple variational image assimilation for accessible micro-elastography

Aleix Boquet-Pujadas, Jean-Christophe Olivo-Marin
Institut Pasteur, Bioimage Analysis Unit, CNRS UMR3691, Paris, France.
E-mail: aboquetp@pasteur.fr, jcolivo@pasteur.fr

Abstract. The mechanical properties of organic materials are relevant to biological processes at many scales. For example, pathological and physiological changes in tissue can often be detected by monitoring their elastic properties. At another scale, the stiffness of the substrate has a direct influence on cell function and differentiation. However, current elastographic methods do not yet cover all spatial scales and require complex and expensive experimental set-ups. In this work, we present a method that could be easily implemented in biology laboratories by using any available fluorescent microscope, regardless of its resolution, and a piezoelectric actuator. To this end, we have developed a variational framework that integrates multiple measurements into a PDE-constrained minimization of a customized image motion descriptor. Preliminary results with a linearly elastic PDE model show that this method is able to extract reliable maps of the rigidity modulus under the noisy imaging conditions characteristic of conventional microscopy and within the deformation ranges of commercially available actuators.

1. Introduction

Variations in the mechanical properties of biological tissue reflect functional changes at the cell scale [1]. This relationship can be exploited to indirectly characterize pathological and physiological changes. For example, at the organ scale, cirrhotic livers become stiffer; at an intermediate scale, tumors are harder than their surrounding tissue; and at the cell scale, the stiffness of myocites is in phase with the cardiac beat [2]. Therefore, an accurate quantification of the elastic parameters within biological materials at several spatial scales is essential to provide good diagnoses.

Precisely, the aim of elastography is to yield stiffness maps of soft tissues by use of biomedical imaging. The most prominent techniques in elastography are ultrasound and magnetic resonance imaging. These methods are focused on extracting tissue features at the macro scale for clinical diagnosis, maxing out at a spatial resolution in the order of the millimeter [3]. However, finer scales are relevant to many biological processes and could also yield more precise diagnostic tools. Respective examples are the influence of the elastic extracellular matrix on cell migration and differentiation, and the amelioration of cancer-recurrence ratios with more precise tumor excisions. Elastography has progressed into the micro scale by turning to optical imaging, including techniques such as optical coherence tomography [4] and Brillouin microscopy [5]. These techniques usually yield strain measurements and have only been recently complemented with indirect stress sensors [6] or shear waves [7] to yield tissue elasticity. At the smallest scale, atomic force microscopy is able to extract measurements of nanometer precision but requires point by point direct mechanical probing with a cantilever. In any case, all these methods require complex and expensive experimental set-ups that depend on precise calibrations and impede widespread adoption, notably for non-clinical research. In this paper, we propose a computational method to extract elasticity measurements in biological materials that only requires (i) a standard microscopy set-up, available in most biology laboratories, and (ii) a piezoelectric actuator [6] or, possibly, a chip with a vacuum chamber [8] (both commercially available). Moreover, the scale of the resulting measurements is directly linked to the resolutive power of the microscope and thus can range across several scales, potentially bridging the gap between the nano and the micro scale [9] with the advent of super-resolution. We note, however, that our system is model-dependent and, even though in theory any type of imaging can be used, in the case of fluorescence microscopy some materials may require a label in the form of either fluorescent proteins or beads.

More concretely, the experimental set-up (Fig. 1a) consists of a piezoelectric (PZT) actuator that moves a glass slide to compress the sample with great precision, either from the top or from the side, generating one or more boundary conditions. If necessary, the sample can be rotated to generate additional measurements. We note that these actuators and set-ups have been used to induce compression at different scales, generating displacements over a wide range of magnitudes within different samples [6], [10]. For example, the PZT ring actuator by *Piezomechanik* in [6] is able to induce up to 50% strain in step-wise micrometer-scale displacements on different synthetic and biological materials. Besides, more expensive models are able to offer nanometer accuracy. The strain induced within the sample can thus be regulated at almost any scale and is usually extracted by taking pictures before and after the compression is applied. These can be taken in a range of resolutions by using ultrasounds, optical coherence tomography or, as in the present case, conventional fluorescence microscopy.

Our technique consists in extracting the internal motion of the sample observed in this way under the different prescribed boundary displacements, while simultaneously inferring the elasticity parameter of a physical model of the sample's material. Namely, we minimize a (custom-modified) optical flow functional [11] describing the movement of the pixel intensity within the images (Sec. 3), constrained by a system of PDEs [12] modeling the relation between the movement of the observed material and its elastic parameter of interest (Sec. 2). The constrained minimization is then linearized and embedded in an image pyramid scheme, and solved by a combination of the Finite Element Method and gradient descent based on the adjoint method (Sec. 5). We exemplify this method with an isotropic linearly elastic model [13]. This is already a good approximation because the displacements are prescribed by the user and because the tissue integrity cannot be compromised.

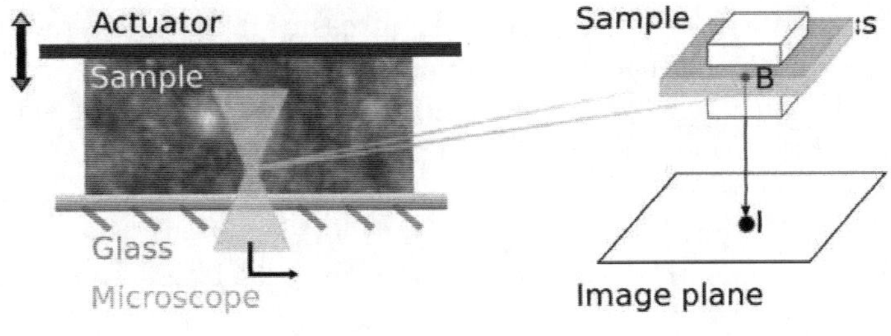

(a) (b)

Figure 1. Schematic of the experimental set-up. **(a)** The compression induced by the actuator on the sample can be finely regulated and can be imaged through a fluorescence microscope. Two images of the sample are taken, one before (cyan) and one after (red) the compression is applied. **(b)** The brightness B in the sample is projected by the laser (with section s) onto the imaging plane as intensity I. The movement of B in 3D is not the same as that of I in 2D.

IOP Conf. Series: Journal of Physics: Conf. Series **1131** (2018) 012014 doi:10.1088/1742-6596/1131/1/012014

The coupled constrained minimization and the integration of several measurements (Sec. 4) make the method able to work with image noise, specially farther from the compression source (where the strain is smaller), attenuate the error propagation of decoupled methods, and offer implicit measurements of the stress. We present results (Sec. 6) on simulations reproducing the real imaging process. That is, on real microscopy images, under real noise conditions, and using real actuator boundary conditions that induce strains as in the literature.

2. The forward problem

We consider our sample to be a linearly elastic material with domain $\Omega \subset \mathbb{R}^d$, $d \in \{2,3\}$, and boundary $\partial\Omega$. Its potential energy $E(\mathbf{u})$ is a function of the displacement vector field $\mathbf{u} : \Omega \mapsto \mathbb{R}^d$:

$$E(\mathbf{u}) := \frac{1}{2}\int_\Omega \boldsymbol{\sigma}(\mathbf{u}) : \boldsymbol{\varepsilon}(\mathbf{u}) - \int_\Omega \mathbf{b}\cdot\mathbf{u} - \int_{\partial\Omega} \mathbf{T}\cdot\mathbf{u}, \tag{1}$$

where : stands for the Frobenius inner product. In more detail, E is a function of two symmetric d^2-dimensional second rank tensors: the strain $\boldsymbol{\varepsilon}(\mathbf{u})$, describing the relative displacement under a deformation; and the stress $\boldsymbol{\sigma}(\boldsymbol{\varepsilon}(\mathbf{u}))$, representing the internal forces within the continuous material caused by the deformation. In our case, we work with small strain theory, $\boldsymbol{\varepsilon}(\mathbf{u}) := 1/2\left(\nabla\mathbf{u} + \nabla\mathbf{u}^T\right)$; and we choose isotropic conditions for the constitutive equation, $\boldsymbol{\sigma}(\mathbf{u}) := \lambda\,\mathrm{tr}\,(\boldsymbol{\varepsilon}(\mathbf{u}))\mathbf{I} + 2\mu\boldsymbol{\varepsilon}(\mathbf{u})$ where $\mu > 0$, $\lambda > 0$ are the Lamé parameters that characterize the material and $\mathbf{I}$ the d-dimensional identity matrix. E is also a function of the body and traction forces $\mathbf{b}$, $\mathbf{T}$. We define the functional derivative of E with respect to $\mathbf{u}$ in the direction $d\mathbf{u}$ as usual:

$$d_{\mathbf{u}}E = \partial_u E \cdot d\mathbf{u} = \left.\frac{dE(\mathbf{u} + \epsilon d\mathbf{u})}{d\epsilon}\right|_{\epsilon=0}. \tag{2}$$

We can then use it to derive the elastic equations of equilibrium from the perspective of energy minimization, $d_u E = 0$, by perturbing the displacement $\mathbf{u}$ with $\mathbf{w} = d\mathbf{u}$. In the present case of linear elasticity, it is enough to use the linearity of operators $\boldsymbol{\varepsilon}$, $\boldsymbol{\sigma}$ and their symmetries to derive:

$$a(\mathbf{u}, \mathbf{w}) = 0 \quad \forall\ \mathbf{w}, \tag{3}$$

$$\text{where } a(\mathbf{u}, \mathbf{w}) := \int_\Omega \boldsymbol{\sigma}(\mathbf{u}) : \boldsymbol{\varepsilon}(\mathbf{w}) - \int_\Omega \mathbf{b}\cdot\mathbf{w} - \int_{\partial\Omega} \mathbf{T}\cdot\mathbf{w}. \tag{4}$$

By choosing the weighting and trial spaces as $\mathbf{u} \in H^1_{\mathbf{g}}(\Omega)$ and $\mathbf{w} \in H^1_{\mathbf{0}}(\Omega)$, where $\mathbf{g}$ and $\mathbf{0}$ are the respective boundary conditions, eq. (3) becomes the standard weak formulation of linear elasticity. We note that we will not consider body forces and that the traction stresses will be weighted out by the Dirichlet conditions, i.e. $\mathbf{u} = \mathbf{g}$ on $\partial\Omega$.

While the framework presented throughout this paper can be used in 3D imaging, this data is still not widely available in biology. Here and in Sec. 3, we adapt it to 2D imaging, which is the condition appropriate to our experiments. Therefore, we consider a plane-stress model, i.e. $\boldsymbol{\sigma}_{i,z} = 0$. In this case, the strain is uniform over the thickness of the sample and the planar components of the displacement $\mathbf{u}_{x,y}$ satisfy eq. (3) in $d = 2$. Thus, we use $\mathbf{u}$ and $\mathbf{x}$ to refer to the planar $d = 2$ case from here on. In addition, we have

$$\nabla\cdot\mathbf{u} = -2\mu\partial_z u_z/\lambda, \tag{5}$$

where u_z is the out-of-plane component. Given a map of the varying rigidity modulus $\mu(\mathbf{x})$ within the sample, $\mathbf{x} \in \Omega$, the forward problem consists in finding $\mathbf{u}(\mathbf{x})$ that satisfies eq. (3).

3. The data term

The problem is to find $\mu(\mathbf{x})$ from the material deformation $\mathbf{u}(\mathbf{x})$. However, this displacement field is not directly available and has to be retrieved from images. To this end, two images are acquired. One before applying a known $\mathbf{g}$ on the sample boundary $\partial\Omega$, and one after. We define $I(\mathbf{x}, t)$ as the function representing the evolution of the image intensity from the first image $I(\mathbf{x}, 0)$ to the second $I(\mathbf{x}, 1)$. Optical flow techniques [14] usually extract the motion $\mathbf{o}$ between the images by (i) assuming the intensity is conserved, i.e. $dI/dt = \mathbf{o} \cdot \nabla I + \partial_t I = 0$, and (ii) imposing that the motion is regular in some sense.

Nonetheless, the motion $\mathbf{o}$ of the intensity within the image does not directly reflect the original motion in the sample described by $\mathbf{u}$ and u_z. For example, material can go out of the 2D imaging plane. To amend this, we consider the whole imaging process, from the emission of light to the final orthographic projection (Fig. 1b). We begin by assuming that it is the original brightness $B(\mathbf{x}, z, t)$ emitted by the fluorophores in the 3D material that is conserved, i.e. $dB/dt = 0$. When a laser is shone, the microscope collects the 3D brightness of the sample within the section s of the laser sheet and projects it orthographically onto the 2D imaging plane [15]. The resulting intensity is proportional to the integral of the brightness: $I(\mathbf{x}, t) \propto \int_s B(\mathbf{x}, z, t)dz$. From this and the conservation equation for the brightness we derive a new conservation equation for the intensity: $dI/dt = \mathbf{v} \cdot \nabla I + \partial_t I = \partial_z u_z I$ up to first order (whereas we would recover $dI/dt = 0$ for a 3D-stack acquisition), where $\mathbf{v} := \int_s B\mathbf{u} / \int_s B$ is the brightness-weighted displacement. Notice that the material derivative of the intensity has gained an extra source term that accounts for out-of-plane movement.

Using the plane-stress assumption and eq. (5), we have $\partial_z \mathbf{u} = 0$, $\mathbf{v} = \mathbf{u}$, and thus the conservation equation is exact and relates the motion $\mathbf{u}$ of the sample directly with the intensity: $dI/dt = -\lambda \nabla \cdot \mathbf{u}/(2\mu)$. This is an ordinary differential equation and can be integrated à la [16], assuming a constant displacement field: $I(\mathbf{x} + \mathbf{u}, 1) = I(\mathbf{x}, 0) \exp(-\lambda \nabla \cdot \mathbf{u}/(2\mu))$. The problem of motion estimation is then to find $\mathbf{u}$ that minimizes

$$D(I, \mathbf{u}; \mu) = \int_\Omega \left(I(\mathbf{x} + \mathbf{u}, 1) - I(\mathbf{x}, 0)\, e^{-\frac{\lambda \nabla \cdot \mathbf{u}}{2\mu}} \right)^2. \tag{6}$$

Since this problem is non-convex, we linearize the inner expression by Taylor expansion. However, the resulting functional is only valid for small displacements of the order of one pixel. Therefore, we embed the problem into a multi-resolution scheme [17], creating a Gaussian pyramid of image pairs (see Sec. 5). The problem is then solved progressively, from the coarsest image pairs of the pyramid (where the biggest displacements are only pixel-sized) to the finest, by warping back the second image at each scale with the displacement solution of the previous scale.

The minimization problem is also not well-posed. To help this, the seminal optical flow is regularized by adding a smoothness constraint to the displacement field. In our case, we can regularize the problem in a more physically sound way:

$$\mathcal{R} = \alpha \int_\Omega \mu^2, \tag{7}$$

where $\alpha > 0$ is the parameter that weights the regularization term, relative to the data term D, in the sum $D + \mathcal{R}$.

4. The inverse problem

Our interest is to find a μ that minimizes $D(I, \mathbf{u}; \mu) + \mathcal{R}$ subject to $a(\mathbf{u}, \mathbf{w}; \mu) = 0 \; \forall \, \mathbf{w} \in H_0^1(\Omega)$. In the incompressible case, i.e. $\lambda \to +\infty$, inverting μ from $\mathbf{u}$ is a well-posed problem for certain boundary conditions [18]. However, several complications arise in practical situations. First, biological materials are often compressible. Second, the high levels of noise and the limited sampling in images make the problem very sensitive. Third, in order to conserve the integrity of the sample, only small displacements can be applied as boundary conditions. As a consequence, the displacement field over the sample can be hard to resolve, specially far from the source and in the presence of noise. For these reasons, we introduce a new functional

$$\mathcal{D} = \sum_{i=1}^{n} D(I_i, \mathbf{u}_i; \mu) \tag{8}$$

that incorporates n observations of the sample under different boundary conditions $\mathbf{g}_i$ that yield n different image pairs $I_i(\mathbf{x}, 0)$, $I_i(\mathbf{x}, 1)$. With this strategy, we aim at completing any lack of information stemming from the above-mentioned difficulties by accumulating independent observations. We also have to introduce the corresponding combined weak elastic equation:

$$\mathcal{A} = \sum_{i=1}^{n} a(\mathbf{u}_i, \mathbf{w}_i; \mu), \tag{9}$$

where $\mathbf{U} = (\mathbf{u}_1, \cdots, \mathbf{u}_n) \in \mathcal{U} := \bigtimes_i H^1_{\mathbf{g}_i}(\Omega)$ and $\mathbf{W} = (u_1, \cdots, u_n) \in \mathcal{W} := \bigtimes_i H^1_{\mathbf{0}}(\Omega)$. The final problem thus reads:

$$\underset{\mu}{\text{argmin}}\ \mathcal{D} + \mathcal{R} \text{ subject to } \mathcal{A} = 0\ \ \forall\ \mathbf{W} \in \mathcal{W}. \tag{10}$$

We use a quasi-Newton method to solve the optimization problem (10) and compute the derivatives with the adjoint method. By using the chain rule to calculate the necessary derivatives with respect to μ,

$$d_\mu(\mathcal{D} + \mathcal{R}) = \partial_{\mathbf{U}}\mathcal{D} \cdot d_\mu\mathbf{U} + \partial_\mu\mathcal{D} + \partial_\mu\mathcal{R}, \tag{11}$$

we can separate partial and total derivatives. While the partial derivatives $\partial_{\mathbf{U}}\mathcal{D}$, $\partial_\mu\mathcal{D}$, $\partial_\mu\mathcal{R}$ are easy to calculate from eqs. (6) and (7) (either analytically or via finite differences), computing the remaining term is more costly. Indeed, note that finding $d_\mu\mathbf{U}$ requires solving

$$\mathcal{A}(d_\mu\mathbf{U}, \mathbf{W}; \mu) = -\partial_\mu\mathcal{A} \tag{12}$$

$\forall\ \mathbf{W} \in \mathcal{W}$, where we use $\partial_{\mathbf{U}}\mathcal{A}(\mathbf{U}, \cdot; \cdot) \cdot d_\mu\mathbf{U} = \mathcal{A}(d_\mu\mathbf{U}, \cdot; \cdot)$ because $a(\mathbf{u}_i, \cdot; \cdot)$ is linear and thus $A(\mathbf{U}, \cdot; \cdot)$ as well. Eventually, equation (12) would have to be solved for each of the many possible directions of μ, thus being slow. As a faster alternative, we propose to use the adjoint method. In this case, we have $\partial_{\mathbf{U}}\mathcal{D} \cdot d_\mu\mathbf{U} = -\mathbf{W}^* \cdot \partial_\mu\mathcal{A}$ where $\mathbf{W}$ is the solution of the system

$$\mathcal{A}(\mathbf{W}, d_\mu\mathbf{U}; \mu) = \partial_{\mathbf{U}}\mathcal{D}^* \tag{13}$$

$\forall\ d\mathbf{U} \in \mathcal{W}$ because $\mathcal{A}^*(W_1, W_2; \mu) = \mathcal{A}(W_1, W_2; \mu)\ \forall\ W_1, W_2 \in \mathcal{W}$ by construction. Using the adjoint method, we require only one solution of (13) (here the right-hand side does not change as opposed to eq. (12)) and we can evaluate the gradient (11) in the different directions of μ at the cost of dot products with the changing partial derivative $\partial_\mu\mathcal{A}$.

5. Numerical implementation

The Galerkin formulation of the problems is obtained by introducing the mixed finite spaces $\mathcal{U}^h \subset \mathcal{U}$ and $\mathcal{W}^h \subset \mathcal{W}$. The resulting discretization is solved via the Finite Element Method (FEM) [19] with mixed P1 Lagrange elements. In practice, the variational problem $\mathcal{A} = 0$ is assembled into its sparse matrix using the Fenics library [19, 20], and the MUMPS library [21]

IOP Conf. Series: Journal of Physics: Conf. Series **1131** (2018) 012014 doi:10.1088/1742-6596/1131/1/012014

is used to solve the linear system in parallel. In order to calculate the gradient of the functional $\mathcal{D}+\mathcal{R}$ with respect to μ, the adjoint system was originally solved by taking the continuous PDE approach shown in (13), however, we are currently using the very convenient discrete approach offered by the dolfin-adjoint library [22, 23]. The minimization of the functional is performed by gradient descent, using the limited memory BFGS algorithm [24], a quasi-Newton method implemented in the PETSc library [25].

Regarding the image processing, the image pyramid $I_i^k(\mathbf{x},0)$, $I_i^k(\mathbf{x},1)$, $k \in \{1 \mathinner{..} m\}$ is created by iteratively applying Gaussian filters and downsampling each original pair of images $I_i(\mathbf{x},0)$ and $I_i(\mathbf{x},1)$, $i \in \{1 \mathinner{..} n\}$; the spatial gradient $\nabla I_i^k(\mathbf{x},1)$ is also calculated. This and the image warping are done using the standard Python libraries scipy and numpy [26, 27]. At each scale of the pyramid a mesh for the FEM is created automatically by using Mshr and CGAL [28].

The final joint algorithm goes as follows. Starting by the coarsest scale and going towards the finest, at each step k do: warp all the i image pairs $I_i^k(\mathbf{x},1)$ with the previous displacement solution $\mathbf{U}^{k-1}$, i.e. $I_i^k(\mathbf{x}+\mathbf{u}_i^{k-1},1)$, and calculate their gradient $\nabla I_i^k(\mathbf{x}+\mathbf{u}_i^{k-1},1)$. Use these images to build the linearized functional $\mathcal{D}^k$. Minimize the functional by performing gradient descent with initial value μ^{k-1} until the convergence criterion is met and a new μ^k is obtained, then do a final solve of the weak equation to obtain $\mathbf{U}^k$ and move on to the next scale.

6. Examples and discussion

In order to test the methodology, we use different synthetic examples derived from real data, to which we apply simulated actuations. This allows us to work with fully controlled conditions. In particular, we use original microscopy images $I_i(\mathbf{x},0)$ of polyacrylamide gels, a well-studied linearly elastic material used widely in biology, and we use real values for the boundary conditions induced by a piezoelectric actuator, generating displacements like those observed in the literature. We also reproduce the noisy conditions of conventional fluorescence microscopy.

We start with a given modulus $\mu(\mathbf{x})$ map (Fig. 2(c)) and solve the forward problem $\mathcal{A}=0$ under different boundary conditions $\mathbf{g}_i$, $i \in \{1 \mathinner{..} n\}$, to obtain the different displacement fields $\mathbf{u}_i$ (Fig. 2(b), $n=1$). Then we use $\mathbf{u}_i$ to warp images $I_i(\mathbf{x},0)$ into images $I_i(\mathbf{x},1)$ (Fig. 2(a)). Finally, we add noise at the level expected from a microscope and subsample the data to obtain the final pairs of images.

We remark that the magnitude of the displacements quickly decreases as one gets farther from the side where the boundary condition is applied, partially due to the high compressibility of the gel ($\lambda \approx 2$ for many biological applications). In fact, the added noise and the sampling amount to an almost complete loss of information at the opposite side (Fig. 2(a), white). In addition, the side of the image where the compression is applied is another source of error because the sample changes size slightly.

In the bottom row of figure 2, the results of the minimization (eq. 10) are presented. Figure 2(d) shows the $\mu(\mathbf{x})$ solution extracted at each scale of the multi-resolution approach: as more scales are considered, the map is refined. Figures 2(f) and 2(e) show a comparison between the extracted $\mu(\mathbf{x})$, $\mathbf{u}(\mathbf{x})$ and the original data in respective figures 2(c) and 2(b). While the magnitude and shape of $\mu(\mathbf{x})$ are extracted correctly, there are some mismatches caused by the small values (relative to the noise) of the displacement field far from the boundary where the compression is applied, and also by the warping of this side of the image. The reconstruction of $\mathbf{u}(\mathbf{x})$ is very precise but is also affected by the same problems (the field is slightly translated upwards with respect to the original).

To alleviate the artifacts introduced by these conditions, we use our framework to combine several measurements of the same sample (Fig. 3). In this case, we work with a more complicated map of the rigidity modulus (Fig. 3(a)), consisting of three inclusions of different magnitude that occlude each other. By integrating additional data in the form of three different boundary conditions, $\mu(\mathbf{x})$ is better recovered (Fig. 3(b)) because the effects from warping the boundary

IOP Conf. Series: Journal of Physics: Conf. Series **1131** (2018) 012014 doi:10.1088/1742-6596/1131/1/012014

are reduced and because the combination of the displacement fields samples the domain more extensively. In this way, all three inclusions are resolved, localized and ordered in magnitude.

Given the fact that we are using real image data, we are very encouraged by our preliminary results. Our method is able to tackle the compressibility of the material and the experimental noise. We believe it fares well as compared to other approaches found in the literature, whether only theoretical (where the inversion is done directly on the displacement field and under less noise, e.g. [13, 29]) or applied (see [6, 30]). Finally, the system will be further validated experimentally by using phantoms with inclusions of known elasticity.

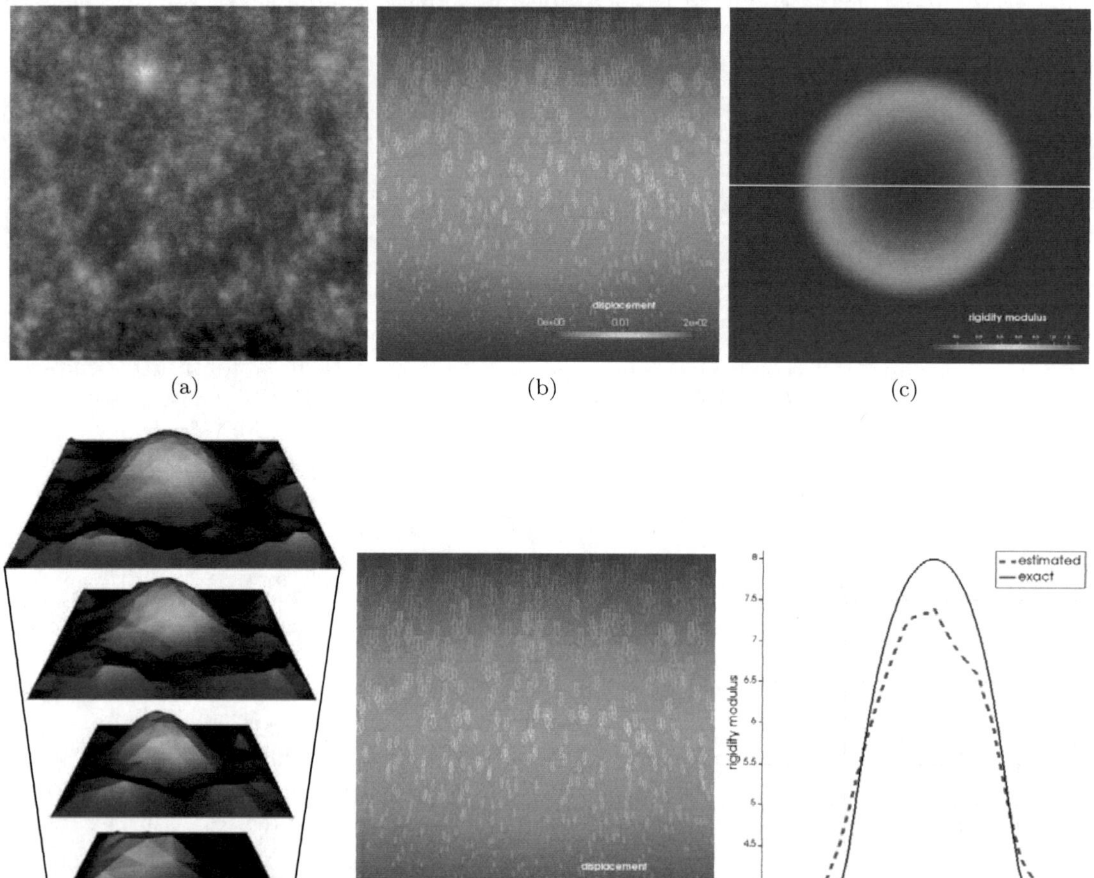

Figure 2. The case of one boundary condition. **(a)** Two superimposed images of the polyacrylamide gel before (cyan) and after (red) applying a uniform top boundary condition downwards and adding noise. Lack of movement is shown in white (cyan+red). **(b)** Displacement field applied to warp (a) resulting from $\mathcal{A}(\mu) = 0$ with the Dirichlet condition. **(c)** Corresponding $\mu(\mathbf{x})$ map used in (b). **(d)** μ solutions of the multi-resolution scheme. **(e)** Recovered displacement field. **(f)** Final μ estimated (dashed line) across the line drawn in (c) as compared to the exact values (solid line).

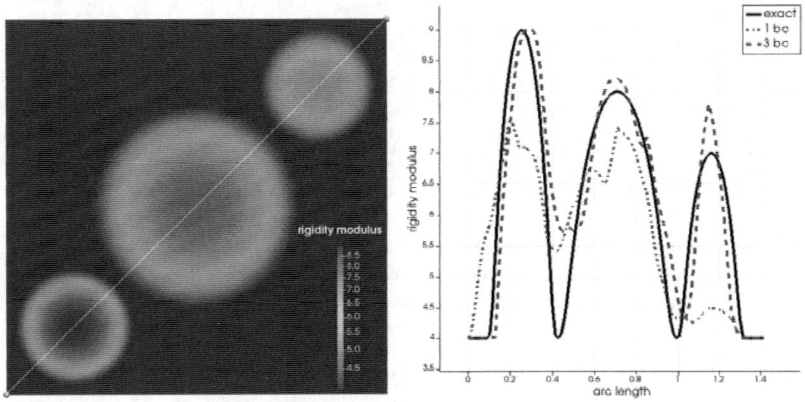

(a) (b)

Figure 3. The case of several boundary conditions. **(a)** Original $\mu(\mathbf{x})$. **(b)** Recovered modulus of rigidity across the diagonal drawn in (a) for 1 and 3 different boundary conditions (dashed lines) as compared to the ground truth (solid line). The respective boundary conditions are compressing from the top, and compressing from the top and the two sides.

Acknowledgments
We acknowledge S. Cuervo for comments on the manuscript, W. Meiniel and V. Meas-Yedid for review, and D. Gonzalez for assistance in image settings. This project has received funding from the European Union's Horizon 2020 research and innovation programme under the Marie Skłodowska-Curie grant agreement No 665807 (Pasteur-Paris University Doctoral Programme) (fellowship to ABP), the Institut Pasteur, the CNRS and the French National Agency for Research (ANR): ANR-10-LABX-62-IBEID and ANR-10-INBS-04-06 France-BioImaging.

References
[1] Y. C. Fung and R. Skalak, "Biomechanics: Mechanical Properties of Living Tissues," *Journal of Biomechanical Engineering*, 1981.
[2] E. U. Azeloglu and K. D. Costa, "Dynamic AFM elastography reveals phase dependent mechanical heterogeneity of beating cardiac myocytes," in *Proceedings of the 31st Annual International Conference of the IEEE Engineering in Medicine and Biology Society: Engineering the Future of Biomedicine, EMBC 2009*, 2009.
[3] R. Righetti, S. Srinivasan, and J. Ophir, "Lateral resolution in elastography," *Ultrasound in Medicine and Biology*, 2003.
[4] J. M. Schmitt, "OCT elastography: imaging microscopic deformation and strain of tissue," *Optics Express*, 1998.
[5] G. Scarcelli and S. H. Yun, "Confocal Brillouin microscopy for three-dimensional mechanical imaging," *Nature Photonics*, 2008.
[6] K. M. Kennedy, L. Chin, R. A. McLaughlin, B. Latham, C. M. Saunders, D. D. Sampson, and B. F. Kennedy, "Quantitative micro-elastography: Imaging of tissue elasticity using compression optical coherence elastography," *Scientific Reports*, 2015.
[7] C. Li, G. Guan, X. Cheng, Z. Huang, and R. K. Wang, "Quantitative elastography provided by surface acoustic waves measured by phase-sensitive optical coherence tomography," *Optics Letters*, 2012.
[8] D. Huh, B. D. Matthews, A. Mammoto, M. Montoya-Zavala, H. Yuan Hsin, and D. E. Ingber, "Reconstituting organ-level lung functions on a chip," *Science*, 2010.
[9] S. Wang and K. V. Larin, "Optical coherence elastography for tissue characterization: A review," 2015.
[10] S. Eshaghian, K. M. Kennedy, P. Gong, D. D. Sampson, R. A. McLaughlin, and B. F. Kennedy, "Optical palpation in vivo: imaging human skin lesions using mechanical contrast," *Journal of Biomedical Optics*, 2015.
[11] B. K. Horn and B. G. Schunck, "Determining optical flow," *Artificial Intelligence*, 1981.

[12] A. Boquet-Pujadas, T. Lecomte, M. Manich, R. Thibeaux, E. Labruyère, N. Guillén, J.-C. Olivo-Marin, and A. C. Dufour, "BioFlow: a non-invasive, image-based method to measure speed, pressure and forces inside living cells," *Scientific Reports*, vol. 7, no. 1, p. 9178, 2017.

[13] A. A. Oberai, N. H. Gokhale, and G. R. Feijóo, "Solution of inverse problems in elasticity imaging using the adjoint method," *Inverse Problems*, 2003.

[14] D. Heitz, E. Mémin, and C. Schnörr, "Variational fluid flow measurements from image sequences: Synopsis and perspectives," *Experiments in Fluids*, 2010.

[15] T. Liu and L. Shen, "Fluid flow and optical flow," *Journal of Fluid Mechanics*, 2008.

[16] T. Corpetti, D. Heitz, G. Arroyo, E. Mémin, and A. Santa-Cruz, "Fluid experimental flow estimation based on an optical-flow scheme," *Experiments in Fluids*, 2006.

[17] N. Papenberg, A. Bruhn, T. Brox, S. Didas, and J. Weickert, "Highly accurate optic flow computation with theoretically justified warping," in *International Journal of Computer Vision*, 2006.

[18] P. E. Barbone and J. C. Bamber, "Quantitative elasticity imaging: What can and cannot be inferred from strain images," *Physics in Medicine and Biology*, 2002.

[19] M. S. Alnaes, J. Blechta, J. Hake, A. Johansson, B. Kehlet, A. Logg, C. Richardson, J. Ring, M. E. Rognes, and G. N. Wells, "The FEniCS Project Version 1.5," *Archive of Numerical Software*, 2015.

[20] A. Logg and G. N. Wells, "DOLFIN," *ACM Transactions on Mathematical Software*, 2010.

[21] P. R. Amestoy, I. S. Duff, J.-Y. L'Excellent, and J. Koster, "A Fully Asynchronous Multifrontal Solver Using Distributed Dynamic Scheduling," *SIAM Journal on Matrix Analysis and Applications*, 2001.

[22] P. E. Farrell, D. A. Ham, S. F. Funke, and M. E. Rognes, "Automated derivation of the adjoint of high-level transient finite element programs," 2012.

[23] S. W. Funke and P. E. Farrell, "A framework for automated PDE-constrained optimisation," *arXiv:1302.3894*, 2013.

[24] R. H. Byrd, P. Lu, J. Nocedal, and C. Zhu, "A Limited Memory Algorithm for Bound Constrained Optimization," *SIAM Journal on Scientific Computing*, 1995.

[25] S. Balay, K. Buschelman, V. Eijkhout, W. Gropp, D. Kaushik, M. Knepley, L. C. Mcinnes, B. Smith, and H. Zhang, "PETSc Users Manual," *ReVision*, 2010.

[26] T. E. Oliphant, "SciPy: Open source scientific tools for Python," *Computing in Science and Engineering*, 2007.

[27] S. Van Der Walt, S. C. Colbert, and G. Varoquaux, "The NumPy array: A structure for efficient numerical computation," *Computing in Science and Engineering*, 2011.

[28] A. Fabri and S. Pion, "CGAL," in *Proceedings of the 17th ACM SIGSPATIAL International Conference on Advances in Geographic Information Systems - GIS '09*, 2009.

[29] P. E. Barbone and A. A. Oberai, "Elastic modulus imaging: Some exact solutions of the compressible elastography inverse problem," *Physics in Medicine and Biology*, 2007.

[30] H. Rivaz, E. M. Boctor, M. A. Choti, and G. D. Hager, "Real-time regularized ultrasound elastography," *IEEE Transactions on Medical Imaging*, 2011.

Institute of Physics
Dirac House, Temple Back
Bristol BS1 6BE UK

ISSN: 1742-6588
ISBN 978-1-5108-7726-9

CURRAN ASSOCIATES INC.
proceedings.com